FRONTIERS IN
BRAIN RESEARCH

FRONTIERS IN BRAIN RESEARCH

EDITED BY JOHN D. FRENCH, M.D.

COLUMBIA UNIVERSITY PRESS

NEW YORK AND LONDON, 1962

THE PAPERS IN THIS VOLUME WERE PRESENTED AT A
SYMPOSIUM HELD AT THE BRAIN RESEARCH INSTITUTE
OF THE UNIVERSITY OF CALIFORNIA, LOS ANGELES,
OCTOBER 14 AND 15, 1961

COPYRIGHT © 1962 BY COLUMBIA UNIVERSITY PRESS

LIBRARY OF CONGRESS CATALOG CARD NUMBER: 62-19908

MANUFACTURED IN THE UNITED STATES OF AMERICA

JOHN D. FRENCH, M.D.

Opening Remarks

THE OPENING CEREMONIES celebrated here today and tomorrow, October 14 and 15, 1961, are gratifying indeed to the 67 investigators who have worked diligently for many years to develop the Brain Research Institute and the physical structure built to house it. These dedicated scientists from fourteen different departments of the University of California, Los Angeles are witnessing here the fruition of their efforts to create an important research environment suitable for implementing their common research interest in the organ responsible for man's highest achievements. Inaugural ceremonies are meant for such satisfying reflections; but they are meant also to recognize and honor particular efforts of key persons without whom successful ventures are impossible. It is appropriate, therefore, at the very beginning of this program to cite particular debts which the BRI is pleased to acknowledge gratefully.

In a very large sense, the development of the Institute is the result of the dedication, enterprise and energy of Professor Horace W. Magoun. Dr. Magoun, professor and first chairman of UCLA's Department of Anatomy, conceived the idea of a brain institute at UCLA, organized the effort to develop it, and accepted the principal respon-

JOHN D. FRENCH, M.D., is Director of the Brain Research Institute, University of California, Los Angeles, California.

sibility for sustaining progress of the proposal through periods of delay and disappointment. The Institute is pleased and honored to acknowledge Dr. Magoun as its most distinguished member.

A description of the Institute is to be included in the published proceedings of these opening ceremonies, and it will be apparent to all who read that account that an organization of such magnitude had to have many friends in high places in order for it to achieve maturity. Among those who labored most diligently in behalf of the Institute is Dean Stafford L. Warren of the School of Medicine. And the proposal has always had the vigorous encouragement of the chief administrative office of the campus, for the Institute was conceived during the tenure of Chancellor Raymond A. Allen, delivered by Chancellor Vern O. Knudsen and encouraged to majority by Chancellor Franklin D. Murphy. President Clark Kerr has constantly been an active friend, and it was he who, in December, 1959, signed the documents establishing the Institute as an official body within the structure of the University. The Regents of the University of California have always displayed an enlightened interest in the proposal leading to its final approval, and a number of individual Regents provided pivotal aid during particularly critical periods before the Brain Research Institute became a reality.

The Brain Research Institute gratefully acknowledges the financial assistance it has received from a number of benefactors in providing housing, facilities installation and support for its activities. Funds to construct the new Institute building were supplied by the University of California, the National Institutes of Health and the Department of Mental Hygiene of the State of California. Equipment costs of considerable magnitude were met by these same sponsors as well as by grants-in-aid funds from over 30 other local, state and national friends, each of whom, because of limitations in time here, we are pleased to recognize individually elsewhere.

It is a pleasure to acknowledge gratefully the assistance of the National Science Foundation in providing grant support for the purpose of presenting this symposium.

Finally, I wish to thank those individuals whose efforts have assured that these opening ceremonies will be a fitting inauguration of the Brain Research Institute. Dr. Charles H. Sawyer and his committee have labored hard and effectively for over a year to complete the myriad of arrangements involved in preparing for such an assembly as this. The crowning achievement of this committee has been to bring together here nine distinguished scientists from the United States, Great Britain, France and Germany whose pre-eminence in fields of interest to the Brain Research Institute has established them as acknowledged spokesmen for the scientific disciplines they represent. We are all most grateful for their graciousness in coming here to participate in the central theme of the ceremonies, a symposium entitled *Frontiers in Brain Research*.

We are very pleased, also, to welcome nearly 300 friends from many parts of the world who have come to celebrate this occasion with us. The presence of these colleagues and the warm greetings of many others who were unable to attend have established an environment ideally suited to the purpose for which this meeting is held.

Many members of the scientific and technical staff of the

Institute have accepted responsibilities for the preparation and conduct of this meeting effectively and with dedication.

Acknowledgments

The following authors and publishers have generously granted permission to reproduce figures: Academic Press Incorporated, *Experimental Neurology;* The Royal Society, *Proceedings of the Royal Society;* The American Physiological Society, *Handbook of Physiology;* American Association for the Advancement of Science, *Science,* and Robert Gaunt *et al.;* Springer-Verlag (Heidelberg), *Deutsche Zeitschrift für Nervenheilkunde;* and Springer-Verlag (Vienna), *Acta Neurochirurgica.* The name of the source accompanies the figure in each case and the citation is given in the References.

Contents

Opening Remarks JOHN D. FRENCH, M.D.	v
Development of Brain Research Institutes HORACE W. MAGOUN	1
The Brain Research Institute at UCLA JOHN D. FRENCH, M.D.	41
Neuroanatomy in Relation to Experimental Neurology WILLIAM F. WINDLE	54
Neurophysiological Frontiers in Brain Research ALFRED E. FESSARD	70
Regional Neurochemistry and Its Application to Brain Function SEYMOUR S. KETY, M.D.	97
Modern Trends in Neuropathology PERCIVAL BAILEY, M.D.	121
Chemical Transmission in the Central Nervous System JOHN H. GADDUM	165
The Development of Neuroendocrinology GEOFFREY W. HARRIS	191
New Aspects of Brain Functions Revealed by Brain Diseases ROLF HASSLER	242

HORACE W. MAGOUN

Development of Brain Research Institutes

THE ESTABLISHMENT of research institutes has been a significant feature of current growth of scientific endeavor in many fields. Prior to their founding, the long-time encouragement of science came chiefly from two sources: the universities and the scientific academies. The universities, dating from the medieval period, were traditionally concerned with the preservation of knowledge and its transmission in education, but they supported investigation as well. The scientific academies, appearing with the Enlightenment of the seventeenth and eighteenth centuries, were primarily committed to the acquisition of knowledge and therefore naturally played an influential role in promoting research. The first widespread appearance of research institutes toward the end of the nineteenth century was largely attributable, however, to practical advances in immunotherapy and to their contributions to public health.

In Germany, the discoveries of Robert Koch were recognized by the establishment, in 1880, in Berlin, of an Institute for Infectious Diseases. Pasteur's discoveries and especially his demonstration of a preventive treatment for

H. W. MAGOUN is Professor and member of the Department of Anatomy, Brain Research Institute, University of California, Los Angeles, California.

rabies led the French Academy to open an international subscription for the establishment of a Pasteur Institute, ceremoniously inaugurated in Paris in 1888. Other centers established in this pattern included the Institute for Experimental Medicine, founded in St. Petersburg in 1890; the Institute of Preventive Medicine, founded in London in 1891, with Lister as its first chairman; and the Rockefeller Institute for Medical Research, founded in New York in 1901, as the first large-scale research institute to be established in the United States (4).

By the turn of the century, these developments had brought the concept of research institutes to widespread public attention and had demonstrated the great benefit to investigative activity which their establishment made possible. With infectious disease and immunology obviously well provided for, it was natural that succeeding efforts should be devoted to comparable developments in other fields. It was at this point that institutes for brain research, a number of which had been developing locally through the initiative of individual investigators, first received general recognition and support.

At an inaugural assembly of the International Association of Academies held in Paris in 1901, the Royal Academy of Sciences of Saxony formally moved to encourage increased research upon the brain. This action was initiated by Wilhelm His, Professor of Anatomy at Leipzig, who had become impressed by the advantages of research institutes on a visit to the zoological station in Naples in 1886 (16). The Association sanctioned the establishment of an International Brain Research Commission dedicated to promoting study of the structure and function of the central nervous system. This Brain Commission hoped, through the influence of the academies, to convince the various governments of the importance of increasing international activities in this field, primarily by establishing central institutes for brain research where special collections of investigative material could be accumulated and facilities made available for their study.

The members of this International Brain Research Commission were chosen by nomination from learned societies of the different countries, with the Royal Society of London acting as "teller." A small central committee consisting of His, Waldeyer, Ehlers, Munk, Flechsig, Golgi and Obersteiner was formed to facilitate action. His (17) had planned the committee but died before its first meeting, and Waldeyer succeeded him as chairman. By 1907, sixteen different countries were represented in the Brain Commission, whose membership then numbered forty (33).* Seven special committees with heavy anatomical emphasis were formed to coordinate aspects of brain research dealing with descriptive anatomy (Waldeyer, chairman), comparative anatomy (Ehlers), histology (Golgi), embryology (Retzius), pathological anatomy (Obersteiner), physiology (Munk) and clinical neurology (Flechsig). By 1908, seven already existing research institutes had been made associates of the International Brain Research Commission (34) and later some of them were designated as Cen-

* Elliot Smith, Cairo; Gehucten, Louvain; Hansen, Copenhagen; Edinger, Frankfurt am Main; Ehlers, Göttingen; Flechsig, Leipzig; Munk and Waldeyer, Berlin; Langley, Cambridge; Sherrington, Liverpool; Horsley and Ferrier, London; Cunningham, Edinburgh; Dejerine, Raymond, Girard, Lannelongue and Manouvier, Paris; Kure, Tokyo; Golgi, Pavia; Luciana and Mingazzini, Rome; Romiti, Pisa; Mosso and Lugaro, Turin; Winkler, Amsterdam; Guldberg, Christiania; Exner, Obersteiner and Zuckerkandl, Vienna; Lenhossek, Budapest; Bechterev and Dogiel, St. Petersburg; Henschen and Retzius, Stockholm; Monakow, Zurich; Cajal, Madrid; Mall, Baltimore; Minot, Boston; and Donaldson, Philadelphia.

tral Institutes: the Neurological Institute of the University of Vienna (Obersteiner), the Neurohistological Institute of Madrid (Cajal), the Senckenberg Neurological Institute of Frankfurt am Main (Edinger), the Institute of the Neurological-Psychiatric Clinic at Leipzig (Flechsig), the Neurological Institute at Zurich (von Monakow), the Psycho-Neurological Institute at St. Petersburg (Bechterev) and the Wistar Institute of Anatomy and Biology at Philadelphia (Donaldson and Greenman). In 1909, an eighth, the Netherlands Central Institute for Brain Research (Kappers), was newly established at Amsterdam. These remarkable achievements of the Brain Commission ceased with World War I.

As a background for consideration of each of these centers, it may be pointed out that with the development of sectioning and staining procedures in the later nineteenth century, it first became possible to explore the internal organization of man's brain, so important both for understanding its function and for diagnosis and therapy in clinical neurology and psychiatry. With little more than the previous foundation in descriptive external anatomy and gross dissection, the laboratories and investigators of the turn of the century pursued the study of the internal structure of the central nervous system by a variety of methods. Because the brain of the normal adult man was so bewilderingly complex, attention was directed to individuals with special endowments, to clinical cases with focal injury or special pathology, to instances of teratology, as well as to simpler embryological stages, comparative anatomical specimens and animal brains with regional degeneration following experimental procedures.

When stained with differential methods, these provided a range of material for ubiquitous microscopic examination. With this growth of multidisciplinary study, it became difficult for single departments to support laboratories with adequate resources.

A few zealous and enterprising individuals had earlier developed centers, however, in which more or less the whole scope of investigative approach was represented. Among these were the laboratories of Obersteiner in Vienna, of Edinger in Frankfurt, of von Monakow in Zurich, of the Vogts in Berlin and of Flechsig in Leipzig. All were established privately and for a long time were supported chiefly or entirely by income from the neuropsychiatric practice of each director. With the advent of the International Brain Research Commission and its plan for associated institutes, the status of these laboratories was much improved. While previously occupying crowded and often borrowed space, they were accorded more extensive facilities and were usually assigned a supporting budget and university affiliation as well. The interest and support of the International Brain Research Commission thus led to great improvement in the resources and programs of the institutes, each of which may now be considered briefly.

The first neurological institute was established in Vienna in 1882 by Heinrich Obersteiner, who had graduated in medicine there and had undertaken research in von Brücke's laboratory while still an undergraduate (20, 21, 24). Beginning in two rooms in the old Physiological Institute, his laboratory gradually increased, became a part of the University of Vienna in 1905 and, in 1907, on being designated the Central Institute for Brain Research in Austria, moved into newly constructed quarters. At this time its library contained 60,000 items, of which Obersteiner, a great bibliophile, had contributed more than half. His versatile mind encompassed nearly every aspect of neurology and psychiatry, but the research activity of the Institute was centered chiefly on neuroanatomy and pathology. Contributions to these fields were published in the *Arbeiten aus dem Neurologische Institut am Universität Wien,* begun in 1892. In 1938, the Institute was revived by Hans Hoff and Franz Seitelberger; the latter is currently its director. As a measure of its tenure, it may be noted that the 75th Anniversary of the Vienna Institute was celebrated in 1957.

During the 1880's a Brain Research Institute was established by Paul Flechsig in relation to the Neurological-Psychiatric Clinic at the University of Leipzig, where he was Professor of Psychiatry (31, 14). Flechsig contributed importantly to the understanding of the organization of the cerebral cortex through myelogenetic and other studies, in which he was joined by such students as Beevor, Bechterev, Darkschevitsch, Schutz, Vogt and Held. Both for his own contributions and as an associate of His at Leipzig, Flechsig was made an initial member of the Central Committee of the Brain Commission. Upon the designation of his laboratory as one of its associate institutes (34), he reported in 1908 that it had

> begun a collection of brains from persons known during life, from which suitable cases will be described. In addition, there is a large pathological-anatomical collection of the brains of persons suffering from speech defects, and also brains of anthropoids, among which are two gorilla brains; also a collec-

tion of 60,000 brain sections. Finally there is a department for experimental psychology and one for chemical work.

Activities of the Brain-Anatomical Institute in Zurich began in 1886, when Constantin von Monakow was joined by a young American zoologist, Henry Donaldson. In a room borrowed from the Pathological Institute, they cut and studied sections of the brain of a dog from which Hermann Munk, in Berlin, had extirpated both occipital lobes with resulting cortical blindness (22, 23, 32). As in the case of Vienna, where the work of Turck and Meynert preceded the establishment of a neurological institute by Obersteiner, so in Zurich Goll had described the posterior column bundle which bears his name; while von Gudden, Hitzig and Forel succeeded one another at the Psychiatric University Clinic before von Monakow established a brain anatomy institute.

From 1905 to 1916, the *Arbeiten aus dem Hirnanatomischen Institut Zurich* reported the numerous contributions of von Monakow and his students, Fuse, Tsuchida and others. Initial studies were concerned with organization of the visual and acoustic pathways and with thalamo-cortical relations. Subsequently, von Monakow introduced dynamic concepts of diaschisis to account for functional impairment following injury to the brain, and in the last part of his career, he attempted to relate ethical and philosophical values to his earlier neurobiological work. His Institute became part of the University of Zurich in 1910 and, in 1928, von Monakow was succeeded as director by M. Minkowski, whose research interests lay in the maturation of motor function systematically compared with the development of the fetal central nervous system. In 1961,

Konrad Akert was appointed director of the Institute of Brain Research in which anatomical studies are related to programs of neurophysiology, initiated at Zurich in the 1930's and 40's by W. R. Hess and his students.

In 1901, a considerable time after these programs of brain research had been established in Vienna, Leipzig and Zurich, the Spanish government recognized the achievements of Santiago Ramón y Cajal, Professor of Anatomy at the University of Madrid, by creating a Laboratory of Biological Research, called the Instituto Cajal (4, 28). It occupied the third floor of a small building next to the university museum, but the expanded facilities provided great impetus to Cajal, who, in 1888, identified the nerve cell as the unit of neural structure and provided a monumental amount of novel information concerning the histological organization of the brain. In 1906 Cajal shared the Nobel Prize with Golgi, whose reduced silver stain he had put to such profitable use. Upon Cajal's retirement at the age of 70 from the University of Madrid, the Spanish government established a large new Instituto Cajal, completed in 1932, whose research programs, interrupted by the Spanish Civil War, are currently being restored under the directorship of Julian Sanz Ibanez.

The Senckenberg Neurological Institute at Frankfurt am Main was established by Ludwig Edinger (1, 7, 9, 18, 19). In addition to practicing clinical neurology in Frankfurt, Edinger pursued myelogenetic and other neuroanatomical studies, first in a laboratory in his mother's home and, from 1902, in a borrowed room in Weigert's pathological laboratory. Following Weigert's death in 1904, the Senckenberg Foundation constructed a neurological institute on the upper floor of a new building and appointed

Edinger director of its programs in comparative neurology and neuropathology. Its financial support was derived from Edinger's clinical practice, and he contributed also a library of 6,000 volumes on neuroanatomy and extensive collections of brain sections from amphioxus to man. In 1914 this neurological institute was made part of the newly founded University of Frankfurt.

With an interest in evolution carried over from his early education, Edinger sought an understanding of the functional organization of the human brain in terms of comparative neuroanatomy. Already in 1886 he had formed the idea that "perhaps the brain consisted of two different parts, one for elementary functions and another which develops along with the evolution of the animal series." His later work confirmed this view and permitted differentiation of the paleoencephalon, or stem of the brain—which does not change greatly in evolution—from neoencephalic structures, consisting of the cerebral and cerebellar hemispheres—which grow in substance with vertebrate evolution and achieve highest development in man.

The Psycho-Neurological Institute at St. Petersburg was established and directed by V. Bechterev who, following graduation from its Military Medical Academy, spent a period in Europe working with Flechsig and Wundt in Leipzig (35). Returning to Russia in 1886 as Professor of Psychiatry at the University of Kazan, Bechterev "organized a psychiatric clinic, the first psycho-physiological laboratory in Russia, and the first 'Brain Institute' in which the anatomy and physiology of the nervous system were investigated in relevant connection with clinical experience."

In 1893 Bechterev was awarded the Chair of Nervous and Mental Diseases at the Military Medical Academy of St. Petersburg, which was then the foremost academic center in Russia. "In 1907, with the help of private funds, he founded and became Director of the Psycho-Neurological Institute at St. Petersburg. According to his scale of plans, the Institute was to embody his ideas of liberal education in medicine, psychology and sociology, and to serve as a research center for the study of behavior in the broadest sense of this broad term—neuroanatomical, physiological, psychological and sociological, with the essential qualification—objectively." After 1920, this Psycho-Neurological Institute became the State Institute of Brain Research in Leningrad.

In addition to improving the status of existing institutes, the influence of the International Brain Research Commission was instrumental in establishing, in 1909, a completely new Central Institute for Brain Research in Amsterdam (2, 5). Its director, C. U. Ariëns-Kappers, was attracted from Edinger's Institute in Frankfurt where he was already directing the program in comparative neuroanatomy. In an inaugural address, Ariëns-Kappers outlined the search for general principles of neural organization, which was to characterize the chief interest of the Dutch Institute, as it did that of all the others of the time. Advantage was to be taken of the gradual development of the brain with emphasis upon ontogenetic and comparative studies. "More and more," Kappers pointed out, "the idea will urge itself upon us that there is no nerve cell whose place is not determined by fixed law, no dendrite growing in a certain direction and no tract whose origin and ending are not prescribed by fixed rules." To

Kappers, the concept of neurobiotaxis accounted best for the basic determinism in brain organization, which he pursued through early work on phylogenetic displacement of the motor ganglia of the oblongata, which he rather lyrically described as the "quadrille of the nuclei," to later preoccupation with endocranial casts of Peking man and racial migrations in near-Eastern prehistory, cultivated during visiting appointments at Peking and Beirut in the latter part of his career. The fiftieth anniversary of the Netherlands Central Institute for Brain Research was celebrated in 1959, with an International Symposium on "The Structure and Function of the Cerebral Cortex" which marks the direction of its current investigative program under the directorship of S. T. Bok.

During the period when the plans of the Brain Commission were being formulated, an independent development in Berlin led ultimately to the establishment of another major institute for brain research (12, 13, 25). The responsible individuals were Oskar Vogt and his wife, Cécile, who had met in Paris during their study of clinical neurology. Returning to Berlin in 1898, they opened the Neurological Central Station which, in 1902, became the Neurobiological Laboratory of the University of Berlin. So great was the Vogts' progress that, in 1931, the Kaiser Wilhelm Institute for Brain Research was opened in Buch, a suburb of Berlin, with support from governmental, municipal and private sources. The Institute was organized into departments for each investigative discipline, and the large staff included Brodmann and Bielschowsky in histology and Kornmüller and Tönnies in electrophysiology.

From an initial interest in brain-mind relationships, the

research program of the Vogts and their colleagues became predominantly concerned with increasingly sharply defined localization of function in the central nervous system. Topistic units were identified both in cyto- and myeloarchitectural study, as well as in responses to stimulation of excitable cortical areas. Focal patterns were similarly recognized in the electroencephalogram, and topistic vulnerability, called "pathoclisis," involving glial and vascular as well as neuronal susceptibility, was proposed in brain injury and disease. A large number of specific fields were ultimately differentiated in the cerebral cortex, and studies were extended to the basal ganglia and thalamus, as well as to the limbic and reticular parts of the brain by the Vogts themselves and by Maximilian Rose, Lorente de Nó, and Olszewski, who were among their many students. In 1937 the Vogts were forced to leave Berlin and, moving to Neustadt in the Schwarzwald, they established another Institute for Brain Research, where work was continued on their extensive neurological collections. Cécile's 75th and Oskar's 80th birthdays were celebrated widely in 1950.

A clear-cut tendency for brain research institutes to reproduce themselves, already attested by derivation of the Dutch institute from that of Edinger at Frankfurt, is even more strikingly illustrated by the multiple offspring of the Vogts' Institute. In Germany, the Max Planck Society is currently establishing a multidepartmental Institut für Hirnforschung, sections of which will be located in different cities. In addition to the Institute at Neustadt, comparative anatomical studies are proceeding in Giessen and behavioral research is under way at Seewiesen. Sections for neurophysiology are being established in Göt-

DEVELOPMENT OF RESEARCH INSTITUTES

tingen and in Munich, for genetics and neurochemistry in Marburg and for tumors and general neurology at Köln. Most interestingly, because of its relation to the old Edinger Institute, still another section of the Max Planck Institut für Hirnforschung, to be opened soon at Frankfurt, will devote itself similarly to neuroanatomy and pathology.

An earlier offspring of the Berlin-Buch Institute was the Brain Research Institute in Moscow, established under Vogt's direction following the death of Lenin in 1924. When Vogt was called to the USSR and commissioned to arrange facilities for studying the Soviet leader's brain, the mansion of a former industrialist was adapted, a group of Soviet scientists was trained and, in the course of two and one-half years, Lenin's brain was cut in complete serial section. In a preliminary report of the study, Vogt (32) called attention to the hypertrophy of pyramidal cells in the association cortex and, drawing analogy to the muscular hypertrophy resulting from repeated physical exercise, suggested that Lenin might be considered an "association-athlete."

The Moscow Brain Institute gained independent status in 1929 and today includes a number of research departments, as well as a museum where dissections of all stages of animal life provide an impressive display of neural evolution. An atlas of cortical cytoarchitecture of the human brain has recently been published by its director, S. A. Sarkisov; electrophysiological work is exploring the activity of individual cortical laminae; and electron microscopical study is currently being introduced.

The manner in which the participation of the United States developed in the programs of the International

Brain Research Commission is of considerable interest. As one of His' most distinguished students, Franklin P. Mall, Professor of Anatomy at Johns Hopkins, was appointed an initial member of the Brain Commission (29). In 1904, when Charles Minot, Professor of Anatomy at Harvard, was asked to join, Mall wrote to him:

Don't for a moment hesitate to join the *Hirnforschung Institut*. It gives you a great opportunity. 1. It will be easier for you to get human embryos and workers; 2. It gives the best possible backing abroad to get funds in America; 3. It puts us in closer touch with foreign workers.

A year later, in 1905, Mall was consulted by Piersol, Professor of Anatomy at Pennsylvania, about the reorganization of the Wistar Institute for Anatomy and Biology in Philadelphia (10, 11). This had originated from the activities of Caspar Wistar, who occupied the Chair of Anatomy at the University of Pennsylvania early in the nineteenth century. Following his death, a quantity of his anatomical preparations was presented to the University and ultimately gained the status of a museum. In 1892 the University of Pennsylvania succeeded in interesting a descendent, General Isaac Wistar, in the incorporation of this museum as a Wistar Institute of Anatomy and Biology. The pursuit of research and the support of publications were introduced in 1905, a step plainly supported by the General, who wrote: "I fully agree that the Wistar Institute should be designed for the use of investigators, rather than a mere gaping public."

It was at this point that Mall, consulted by Piersol, immediately seized the opportunity to promote the plans of the International Brain Commission, which had requested the National Academy to establish a special institute for study of the brain. Mall wrote:

It seems to me that it would be natural and proper that the Wistar Institute should be designated. Through it, all the work in neurology in America could be enlarged and correlated with the work abroad. . . . What I have just written shows, it appears to me, the brilliant opportunity before the Wistar Institute . . . it would be almost criminal if the Wistar Institute did not take part in it.

As a consequence, the Wistar Institute did at that time initiate a neurological program and, on Mall's recommendation, appointed Henry H. Donaldson as its head. Donaldson gathered a large quantity of data relating to the growth of the nervous system, and his interest in the establishment of biological norms also found expression in standardization of the Wistar albino rat as a common laboratory animal. His successor, E. G. Coghill, later introduced research correlating the development of behavior of the salamander with the maturation of its nervous system, but increasingly strained administrative relations led finally to the termination of Coghill's appointment (15).

In 1919 a second effort to establish a major institute for brain research in the United States met with even less success (8). In an address at the opening ceremonies of the Montreal Neurological Institute in 1934, Harvey Cushing (6) recalled:

As World War I drew to a close, a small group of overseas medical officers, whose official positions had thrown them close together, found themselves disinclined to return to their former humdrum professorial tasks. In talking the matter over, they conceived the idea of founding a National Institute of Neurology, whose primary purpose was to aid the government in supervising the further treatment of the disorders and injuries of the nervous system sustained by our soldiers. . . . We looked forward, not only to having suitable wards for organic, psychopathic, infectious and neurosurgical disorders, but also a well-equipped operating suite, proper laboratories for neuropathological investigation and experimentation, a working library and a new organ of publication. . . . It was our ambition to have the organization grow into a postgraduate school for those whose interests pointed toward neurology or any one of its many bypaths. . . . Were such an institute to be put in operation and kept out of politics, we on our part, as whole-time servants, freed from the distractions of private practice, agreed to devote the remainder of our working lives in the effort to make it a success. . . . Neurosurgery was to be my province, and Lewis H. Weed, who had been in charge of an experimental laboratory for the study of nervous diseases under Army auspices, had agreed to become Director of Laboratories.

In retrospect, the government might have been saved some of the hundreds of millions of dollars that have since been expended largely on the care of these very patients, but when it was suggested that the erection and maintenance of such a supervising institute as we had in mind might cost ten million dollars, this was looked upon as fantastic. We then appealed to the Rockefeller Foundation, where we had a warmer reception, and for a time it seemed that the program might be put through. Unhappily, to make the story short, we met with opposition from certain influential quarters; the undertaking finally was abandoned, and we one and all drifted back to our former academic positions. Disappointed as we were, I like to think the seed did not fall on wholly barren ground, and that our long forgotten project may have eased the way for McGill to establish this unit which, let us hope, will set an example to be emulated by large university centers elsewhere.

With support from the Rockefeller Foundation, this Montreal Neurological Institute was formally opened in 1934, six years after its director, Wilder Penfield, had joined McGill University to introduce clinical neurosurgery and neurological research at its medical school

DEVELOPMENT OF RESEARCH INSTITUTES

(26, 27). At the opening ceremony Penfield stated, "We have carved in stone on the outside of this building a simple declaration of our cause in the words, 'Dedicated to the relief of sickness and to the study of neurology.' " In 1953, at the opening of the McConnel Wing, doubling the institute's facilities, he remarked, "From the day of the First Foundation of 19 years ago, this building has housed two activities: 1) a neurological hospital; and 2) a scientific unit, each supported by a separate budget. This is what a clinical institute is—a place of treating patients and a place nearby for the study of the human and scientific problems thus presented."

High motivation contributed importantly to the accomplishments of the Montreal Neurological Institute. In Penfield's words: "The task to which this institute is dedicated in all humbleness of spirit is the achievement of greater understanding of the neurological ills to which man is heir, so that physicians may come to the bedside with healing in their hands." The provision of research laboratories was also significant, for, as Penfield remarked:

No one could become familiar with neurological problems and with the handicaps imposed upon the workers in this field without concluding that adequate laboratories . . . were essential to further progress. For any constructive ideas that I may have contributed I have to thank the guiding influence of the men under whom it has been my good fortune to work. Most important was the initial influence of Sir Charles Sherrington. In his laboratory at Oxford, to search for the hidden truths of neurology became a habit of mind, a coloring to all one's thought.

Additionally, provision was made for research fellows, of whom Penfield said:

The important work of research is carried on by men who have finished all preliminary training and who have not yet undertaken positions of responsibility. Some of them are voluntary laboratory assistants, others are on a research stipend. These men may be wandering students from any part of the world. If well chosen, the research fellows are the most important part of the staff of an institute of this sort.

At the Montreal Neurological Institute investigative activity has flourished—in neuropathology with Cone, in neurochemistry with Elliott and in neurophysiology with Jasper. In Penfield's hands, the neurosurgical operating room was additionally a laboratory and indeed the Institute's program, which over the years contributed so productively to knowledge of the brain, was undertaken jointly in the operating room and laboratories in the search for mechanisms related to epilepsy. In discussing the future Penfield concluded, "Permanence requires that there shall be transmission, through a succession of native sons, of contagious scientific enthusiasm." Appropriately, Theodore C. Rasmussen was recently called back to the Institute's direction as Penfield's successor.

More recently, Cushing's earlier plan to form a national neurological institute has again borne fruit, this time in his own country and with federal support. Following World War II, the United States Public Health Service established the National Institutes of Health in Bethesda, outside Washington, D. C. A multidisciplinary program of neurological research was established jointly by the National Institutes of Mental Health and National Institute of Neurological Diseases and Blindness, under the research directorship of Seymour Kety and his successor,

Robert Livingston. Among its extensive activities have been those in physiology with Marshall, Tasaki, Frank, Lilly and MacLean; in psychology with Rosvold; in pharmacology with Cantoni; in anatomy with Windle; and in biochemistry with Tower and Axelrod; to name but a part of its large program.

More recently, institutes for brain research have developed in this country in university settings as well. An Institute of Neurology was established at Northwestern University Medical School in 1928, with Stephen W. Ranson as director. With limited support, a productive investigative program was maintained until Ranson's death in 1942, with the subsequent closure of the Institute. At this time, another institute was opened in Chicago as the Neuropsychiatric Institute of the University of Illinois Medical School. In it, sections housing neurology, neurosurgery and psychiatry were connected by a basement laboratory, in which similarly productive programs of research were developed by Warren McCulloch, Percival Bailey, Gerhardt Von Bonin and their associates.

This account, which is essentially devoted to pioneer developments, primarily discusses such activities in other countries. I have made reference to only two American Institutes—one at Northwestern University, the other at the University of Illinois—because, through the association of John D. French and myself with them, they form in a sense the parent organizations from which has stemmed the currently opening Brain Research Institute of the University of California, Los Angeles. Its heritage is an exceedingly rich one; may its accomplishments come to occupy a place of merit among those of such a distinguished past.

REFERENCES

1. Ariëns-Kappers, C. U., Ludwig Edinger (1855–1915), Folia Neurobiologica, 9:343–66, 1915.
2. ——, Dutch Central Institute for Brain Research at Amsterdam, in *Methods and Problems of Medical Education*, 10th series, pp. 1–6. New York, Rockefeller Foundation, 1928.
3. Blake, J. B., Scientific institutions since the Renaissance: their role in medical research, Proc. Am. Philos. Soc., 101:31–62, 1957.
4. Cannon, D. F., *Explorer of the Human Brain; the Life of Santiago Ramón y Cajal (1852–1934)*. New York, Schuman, 1949.
5. Crosby, E. C., Address in memory of Cornelius Ubbo Ariëns-Kappers, in Tower, D. B., and J. P. Schadé, eds., *Structure and Function of the Cerebral Cortex*, pp. 1–6. Amsterdam, Elsevier, 1960.
6. Cushing, H., Psychiatrists, neurologists and the neurosurgeon, in *Neurological Biographies and Addresses*, pp. 17–36. London, Oxford Univ. Press, 1936.
7. Edinger, L., Bericht über das Dr. Senckenbergische Neurologische Institute 1902–1906, Frank. Zeit. f. Path., 1:200–04, 1907.
8. Fulton, J. F., *Harvey Cushing, a Biography*. Springfield, Thomas, 1946.
9. Goldstein, K., Ludwig Edinger (1855–1918), Zeit. f. d. ges. Neurol. u. Psych., 44:114–49, 1918.
10. Greenman, M. J., Concerning the Wistar Institute of Anatomy, Anat. Rec., 1:119–24, 1907.
11. ——, The Wistar Institute of Anatomy and Biology, appendix in *Autobiography of Isaac Jones Wistar*. New York, Harpers, 1914.
12. Hassler, R., Cécile und Oskar Vogt, in Kolle, K., ed., *Grosse Nervenärzte*, Bd. 2, pp. 45–64. Stuttgart, Thieme, 1959.
13. Haymaker, W., Cécile and Oskar Vogt, on the occasion of her 75th and his 80th birthday, Neurology, 1:179–204, 1951.
14. ——, ed., *The Founders of Neurology*. Springfield, Thomas, 1953.
15. Herrick, C. J., *George Ellett Coghill, Naturalist and Philosopher*. Chicago, Univ. of Chicago Press, 1949.
16. His, W., Über wissenschaftliche Centralanstalten und speciell

über Centralanstalten zur Förderung der Gehirnkenntniss. *K. Sächsische Gesellschaft der Wissenschafter, Leipzig.* Math. Phys. Klasse, 53:413–36, 1901.

17. ——, Protokoll der von der internationalen Association der Akademien ixeder gesetzen Centralkommission für Gehirnforschung. *K. Sächsische Gesellschaft der Wissenschaften, Leipzig.* Math. Phys. Klasse, 56:2–4, 1904.

18. Krücke, W., Ludwig Edinger (1855–1918), in Scholz, W., ed., *50 Jahre Neuropathologie in Deutschland,* pp. 20–33. Stuttgart, Thieme, 1961.

19. Krücke, W., and H. Spatz, Aus den "Erinnerungen" von Ludwig Edinger, in *Ludwig Edinger (1855–1918),* Schrift. Wissensch. Gesellsch. University of Frankfurt, Naturwisschensh., pp. 1–25. Wiesbaden, Steiner, 1959.

20. Marburg, O., Zu Geschichte des Weiner Neurologischen Institutes (Festschift zu Feier des 25 Jährigen Bestandes des Neurologischen Institutes an der Weiner Universität), Arb. a. d. Neurol. Inst. a. d. Weiner Univ., 15:VII–XXIII, 1907.

21. ——, Heinrich Obersteiner, Arb. a. d. Neurol. Inst. a. d. Univ. Wien., 24:V–XXXII, 1923.

22. Minkowski, M., Die Poliklinik für Nervenkranke und das Hirnanatomische Institut, Zürcher Spitalgeschichte von Regierungsrat des Kantons Zurich, pp. 427–73, 1951.

23. ——, Constantin von Monakows Beiträge und impulse zur Entwicklung der neurologischen Grundprobleme des Aufbaus, der Lokalisation und des Abbaus der nervösen Funktionen, Schweiz, Arch. f. Neurol. u. Psych., 74:27–59, 1954.

24. Obersteiner, H., Die internationale Gehirnforschung, Deutsche Revue, 33:77–82, 1908.

25. Olszewski, J., Cécile und Oskar Vogt, Arch. Neurol. and Psychiat., 64:812–22, 1950.

26. Penfield, W., The significance of the Montreal Neurological Institute, in *Neurological Biographies and Addresses,* pp. 37–54. London, Oxford Univ. Press, 1936.

27. ——, The second foundation of the Montreal Neurological Institute, in *Prospect and Retrospect in Neurology,* pp. 23–42. Boston, Little, Brown, 1955.

28. Ramón y Cajal, S., *Recollections of My Life,* trans. by Craigie, E. H., Memoirs Am. Philos. Soc., Vol. 8, Parts 1 and 2, 1937.

29. Sabin, F. R., *Franklin Paine Mall, the Story of a Mind*. Baltimore, Johns Hopkins Press, 1934.
30. Vogt, O., Bericht über die Arbeiten des Moskower Staats Instituts für Hirnforschung, J. f. Psychol. u. Neurol., 40:108–18, 1929.
31. Von Bonin, G., *Some Papers on the Cerebral Cortex*. Springfield, Thomas, 1960.
32. Von Monakow, C., Über Hirnforschungsinstitute und Hirnmuseen, Arb. a. d. Hirnanatomischen Inst. in Zurich., 6:1–27, 1912.
33. Waldeyer, W., Document 1 of the report of the President of the Brain Commission (Br. C), Anat. Rec., 1:181–86, 1907.
34. ——, Report on the present status of the academic institutes for brain study, together with a report of the meetings of the Executive Committee of the Brain Commission held at Berlin, March 14, 1908, Anat. Rec., 2:428–31, 1908.
35. Yakolev, P. J., Bechterev, in Brazier, M. A. B., ed., *The Central Nervous System and Behavior*. New York, Josiah Macy, Jr. Foundation, 1959.

The figures on the following pages have been selected from a much larger collection, prepared as a pictorial exhibit on *The Development of Brain Research Institutes* for display at the opening ceremonies of the Brain Research Institute, University of California, Los Angeles; and now available for loan exhibit through the Biomedical Library, UCLA.

Grateful thanks are expressed to the considerable number of persons who have generously contributed pictorial and other material, and especially to P. Bailey, D. Beheim-Schwarzbach, C. Courville, C. Estable, W. Haymaker, M. Minkowski, E. Oldberg, K. Pateisky, T. Rasmussen, S. A. Sarkisov, J. P. Schadé, F. Seitelberger, and M. Vogt.

As always, it is a pleasure to acknowledge the helpful cooperation which Louise Darling, Biomedical Librarian, has so graciously provided in the preparation of both the exhibit and this paper.

FIGURE 1. Heinrich Obersteiner at about the time he was establishing the first Institute for Brain Research in Vienna. Marburg (21) has pointed out that in order to understand Obersteiner's significance as an investigator and teacher, it is necessary "to summarize briefly the development of theoretical knowledge before his time. While the gross anatomists, Rolando, Reil, Burdach, Sömmering and Vicq d'Azyr had established the structure of the central nervous system in its external contours, its finer anatomy remained closed. As Hyrtl said, 'The anatomy of the inside of the brain is and will probably always remain a book closed with seven seals and written in hieroglyphics in addition.' "

FIGURE 2. The Vienna Neurological Institute had its beginning in 1882 in two rooms in this old Gewehrfabrik, which housed the Physiological Institute from 1854 until 1900. In 1905 Obersteiner's Institute became a part of the University of Vienna, and in 1907 it became the Central Institute for Brain Research in Austria. In a report to the Brain Commission in 1908 Obersteiner (24) stated: "the Neurological Institute at Vienna would this year celebrate the twenty-fifth anniversary of its founding. The plan of the new Institute to be erected in the Schwarzpannierstrasse has been determined upon. All lines of brain study, especially embryological and histological investigations, will be fostered there. The library contains 60,-000 volumes, of which Herr Obersteiner has given 30,000 (including reprints)."

FIGURE 3. During the 1880's a Brain Research Institute was established by Paul Flechsig in relation to the Neurological-Psychiatric clinic at the University of Leipzig, where he was Professor of Psychiatry. Flechsig contributed importantly to knowledge of the organization of the cerebral cortex by myelogenetic and other studies, in which he was joined by such students as Beever, Bechterev, Darkschevitsch, Schutz, Vogt and Held. Both for his own contributions and as an associate of His at Leipzig, Flechsig was made an initial member of the Central Committee of the International Brain Research Commission.

FIGURE 4. In 1902, at the age of 47, Edinger obtained a room in Weigert's Institute at Frankfurt from which his own Institute developed. In 1906 Edinger (7) reported that it possessed a library of 6,000 titles and a collection of 10,000 preparations, representing all lines of brain study and particularly those of comparative neuroanatomy. His students up to that time included Auerbach, Holmes, Streeter, Bing, Goldstein, d'Hollander and Kappers.

FIGURE 5. Cornelius U. Ariëns-Kappers, director of the Netherlands Central Institute for Brain Research from its beginning in 1909 until his death in 1946, was a world leader in the study of the comparative anatomy and evolution of the brain. Before joining Edinger's Institute in 1900, Kappers had completed his medical training in Amsterdam with a dissertation on the paths and centers of the brain of the fish. Continuing his work with unflagging zeal, in a 1907 article on "The phylogenetic displacement of the motor ganglia of the oblongata, their cause and significance," he first elaborated the principle of neurobiotaxis.

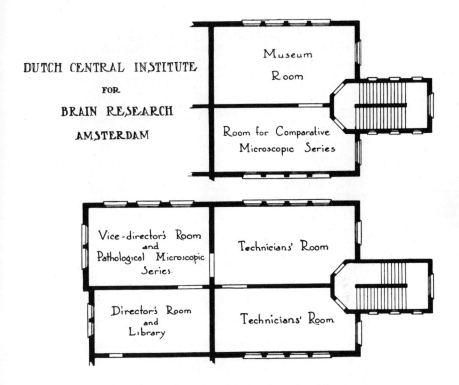

FIGURE 6. In these present days of large-scale construction, it is interesting to note from this floor plan that the Netherlands Central Institute for Brain Research, built in 1908 as a wing of the Anatomical Laboratory, consisted of only six rooms, with space and equipment for four workers. Kappers (2) pointed out that: "Those who work here, including a good many foreigners, are all postgraduates especially interested in fields of science and holding M.D. or Ph.D. degrees. The aid in scientific research work is mostly given by the Director himself, as the Vice-Director is only a half-time official and, moreover, takes care of the museum and the administration." In words that will be familiar to all, he added later: "Though sufficiently large in the beginning, the building has in the course of years become too small."

FIGURE 7. The Brain-Anatomical Institute in Zurich was founded by Constantin von Monakow, who practiced neurology and developed a private research laboratory which, in 1894, became part of the University. Von Monakow held that a special feature of institutes lay in providing opportunities for multidisciplinary study of the brain. In his laboratory, emphasis was laid upon study of cases of focal brain injury collected from the neurological, surgical and psychiatric clinics, as well as from the children's hospital and private practice. In 1912 the collection contained 370 macroscopic and 260 serially sectioned specimens along with 100 series of experimental-anatomical preparations.

FIGURE 8. Bechterev's Psychoneurological Institute was established in St. Petersburg in 1907, both as an institution of higher learning and as a research center comprising several laboratories. Bechterev's research interests were directed to the functional anatomy of the brain, as well as to study of conditional or "associative" reflexes in which somatic rather than visceral motor indicators were employed. He is seen here with two assistants in what appears to have been an underheated laboratory for reflexology. Bechterev's talents for expansive organization found continuing expression in many developments ramifying from his Psychoneurological Institute.

FIGURE 9. Upon graduating in medicine Cajal served in Cuba during the Spanish-American war. With money saved from his military income he purchased a microtome and microscope and began self-taught histological research in an attic laboratory at the University of Zaragoza. On moving to the Chairs of Histology, first at Barcelona and then at Madrid, Cajal perfected Golgi's silver method for neural staining. Employing embryonic material, in which interfering myelin sheaths were undeveloped, he was able to display the structure of the brain with astonishing clarity and made monumental contributions to knowledge of its organization.

FIGURE 10. In 1901 the Spanish government founded the Instituto Cajal on the third floor of this small building (foreground) adjacent to the University of Madrid. Upon Cajal's retirement at 70 the government appropriated support for a large new Instituto Cajal, completed in 1932. "In it," Cajal (28) wrote, "in place of the mean and narrow quarters in which my students work, we shall have at our disposal in the future a magnificent palace not inferior to the proud scientific institutes abroad. There will live together . . . all those among us who are devoted to similar studies. I hope that . . . when they find themselves collaborators in the intellectual renaissance of our country, all will know how to give up our lamentable factionism and individualism, which leads to endless grudges and hard feelings."

FIGURE 11. Cécile and Oskar Vogt met in Paris when she was studying neurology with Pierre Marie and he with the Dejerines. In 1899 they were married in Berlin where Oskar, at the age of 28, had opened a Neurological Central Station, which soon became the Neurobiological Laboratory of the University of Berlin. In 1950 the Vogts celebrated her 75th and his 80th birthday, both working on problems which had occupied their minds for more than half a century.

FIGURE 12. From the beginning Oskar Vogt was fired with the idea of establishing an Institute where scientists of various disciplines could work together on problems related to the nervous system. Both he and Cécile were talented organizers and, in 1931, a large new Kaiser Wilhelm Institute for Brain Research was opened for them in Buch, a suburb of Berlin. The program was organized into a number of departments with a total staff of more than a hundred. Six years later the Vogts were forced to leave Berlin and, moving to the Schwarzwald, they established a second Institute, where their neurological research was continued.

FIGURE 13. This photo shows Professor S. A. Sarkisov, Director of the Brain Institute and Member of the Academy of Medical Sciences, Moscow, USSR. A student of Oskar Vogt, Sarkisov's research interests lie in cortical cytoarchitecture. Initially established for the study of Lenin's brain, the Moscow Brain Institute now includes a multidisciplinary variety of programs directed chiefly to study of the cerebral cortex. It plays an important role in educational and training activities, as well as in research.

FIGURE 14. The Montreal Neurological Institute was formally opened in 1934, six years after its director, Wilder Penfield, seen here (at left) with William Cone and Fellows, had joined McGill University to introduce clinical neurosurgery and neurological research at its Medical School. At the inaugural ceremony Penfield said: "For any constructive ideas that I may have contributed, I have to thank the guiding influence of the men under whom it has been my good fortune to work. Most important was the initial influence of Sir Charles Sherrington. In his laboratory at Oxford, to search for the hidden truths of neurology became a habit of mind, a coloring of all one's thought."

FIGURE 15. The original building (right) was joined by a bridge to the Royal Victoria Hospital and Medical School at McGill. The McConnel Wing (left), added in 1953, doubled the Institute's facilities. Penfield (26) pointed out: "We have carved in stone on the outside of this building a simple declaration of our cause in the words, 'Dedicated to the relief of suffering and to the study of neurology.'" In the investigation of man's brain in the operating room Penfield and his associates have made unprecedented contributions to knowledge of consciousness, speech and memory.

FIGURE 16. The establishment of an Institute of Neurology at Northwestern University Medical School in 1928 enabled its director, S. W. Ranson, to devote full time to research and graduate education and, with limited resources, the Institute's programs flourished until Ranson's death in 1942. Early work was directed to the role of unmyelinated nerve fibers and their central connections in pain conduction. In the second part of the program the study of visceral and endocrine functions led to extensive research upon their higher coordination by the brain stem, in particular the hypothalamus.

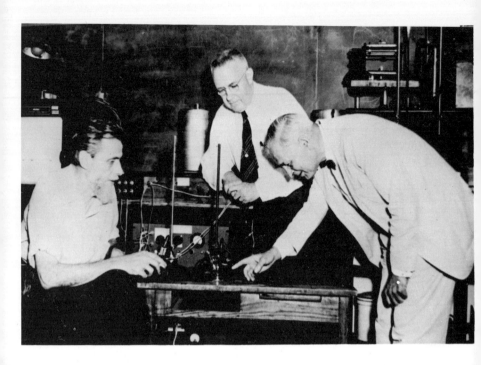

FIGURE 17. The Neuropsychiatric Institute of the University of Illinois was housed in two multi-storied wings connected by a basement research laboratory. Here for a decade Warren McCulloch, Percival Bailey and Gerhardt von Bonin (left to right) collaborated actively in strychnine neuronographic study of cortical interconnections and many other aspects of the function of the brain.

JOHN D. FRENCH, M.D.

The Brain Research Institute at UCLA

THE INAUGURATION of the Brain Research Institute at the University of California, Los Angeles was celebrated officially on October 14 and 15, 1961. This new Institute is the youngest of a long and distinguished list of predecessors of which an historical account has preceded this description. So that friends and colleagues from all over the world may know something about this Institute, it seems appropriate here to give a brief description of its structure and goals in order to identify it among its forebears.

For many reasons, the new Medical School at UCLA, at the time of its founding in 1949, recognized a major interest in the neurological sciences. The 1940's had ushered in a period of increasing enlightenment concerning brain function, attributable largely to advances in the field of neurophysiology but stemming also from developments in the other basic biomedical sciences. Additionally, the nation was becoming acutely conscious of a major responsibility regarding the neuropsychiatric problems of its expanding and aging population. It is not surprising, therefore, that preoccupation with the brain and behavior should emerge as a dominant interest in a medical school founded during that era. This pervasive commitment to brain study led to the development *pari passu* of two sister institutes at UCLA: one was destined to become the Brain Research Institute, dedicated to research in basic disciplines; and the other, the Neuropsychiatric Institute, was designed to implement the clinical management of patients with neurosurgical, neurological and psychiatric disorders. While these two Institutes have complementary interests and extensive collaborative associations, they are administratively quite separate.

Actually, the Brain Research Institute had its inception ten years ago in a small room at the Long Beach Veterans Administration Hospital which Dr. E. V. Edwards, the hospital's manager, was kind enough to allow Dr. H. W. Magoun and me to use as a laboratory. Dr. Magoun had recently come to the new UCLA School of Medicine as Professor of Anatomy and, since the Medical Center was still under construction, off-campus research space had to be developed. The little laboratory at Long Beach, forty miles from the main campus, grew with astonishing rapidity and soon became a full-scale research center. Additionally, it served as a focal point for research planning for the main campus where, through the cordial good will and hard work of a large number of people, programs of basic brain study in most departments came to develop strong interdisciplinary associations. It was this association of programs which, under the leadership of Dr. Magoun, matured into the Brain Research Institute.

The nascent Institute had many problems to solve, but none was more challenging than that of developing laboratory space to house its investigators. A solution was provided in 1958, when the National Institutes of Health agreed to join with the University of California and with the State Department of Mental Hygiene in funding the construction of a ten-story building for the Brain Research Institute. In March, 1961 the building was completed and now it serves to centralize activities of the Institute.

At present, fourteen departments and divisions are represented in the Institute, of which twelve are affiliated with the Medical School (Anatomy, Biophysics, History of Medicine, Infectious Diseases, Neurology, Neurosurgery, Pathology, Pediatrics, Pharmacology, Physiological Chemistry, Physiology, Psychiatry) and two (Psychology and Zoology) are from the College of Letters and Science. In addition, however, the Institute collaborates actively in research programs with many other departments, institutes, laboratories or schools of the University of California as well as with other institutions, particularly the Veterans Administration Hospitals where research programs, developed collaboratively during the formative period of the Institute, continue to be active and productive. Academically and administratively, therefore, the institute is broadly structured, even though it is physically affixed to the Medical Center.

There are 67 member scientists of the Institute, each of whom has an appointment in and responsibilities to participating departments. Membership is awarded in recognition of past accomplishment and future capability in basic research directly related to the neurological sciences. Institute membership does not necessarily convey space assignment, although all senior investigators occupying laboratory facilities within the building are members.

The administration of Institute activities is the responsibility of a director and of an advisory committee of which the director is chairman. All administrative appointments are made by the chancellor of the University with due consideration being given to recommendations made by the director and advisory committee. In general, each department participating in Institute activities is represented on the advisory committee.

While the University of California provides some basic assistance, the bulk of funds required to support the activities of the Institute must be obtained by grants from sources outside the University. Thus, investigators apply for and receive individual project grants to aid the operation of their laboratories. Additionally, however, large grants—now designated by such terms as "base," "program," "block," or even "institutional" awards—have become a necessary and prominent part of the financial structure of the Institute. It is feasible to solicit large sums in the latter categories only from governmental agencies—such as the National Institutes of Health, the National Science Foundation, Atomic Energy Commission, National Aeronautics and Space Administration, The Veterans Administration and the armed services—or from the largest of the private foundations. The broad-based nature of the funding of the Brain Research Institute is indicated by the fact that its budget of several million dollars per year is provided by approximately 160 grants from 20 agencies.

The Brain Research Institute recognizes three principal goals: to undertake basic research in the many fields which contribute to an understanding of brain mechanisms and behavior; to train research investigators for independent careers in brain research; and to develop and disseminate information about brain function in the interests of the social and the scientific communities. Distinction is made among these goals only for purposes of

UCLA'S BRAIN RESEARCH INSTITUTE 45

description since, actually, each is closely related to or overlaps the others.

Basic research is the principal task of the Brain Research Institute. In the early days of the Medical School, studies were sometimes pursued in isolation unintentionally imposed by departmental specialization. As the Institute developed, however, interdepartmental contact was encouraged and the research opportunities which lie between disciplines began to appear. These opportunities were attached collaboratively with each partner in a team bringing special skills to communal research tasks. In addition, investigators found that they were learning new skills from these interdisciplinary exposures and that their own independent research capabilities were becoming extended. Pooling of intellectual and physical resources made it possible to attack many problems too extensive for treatment by individual investigators.

It is feasible here only to illustrate, rather than enumerate, interdepartmental attack upon problems susceptible to "generalist" as well as "specialist" treatment. A Brain Research Institute program of recent origin and considerable importance is one which undertakes to utilize techniques of applied mathematics, communications engineering and computer analysis in resolving problems of brain study. This study recognizes the current developments in these fields which render obsolete older methods of treating such data as EEG tracings or oscilloscopic recordings. Mathematicians, engineers and computer specialists join with neurophysiologists in attacking these problems collaboratively by employing skills of the physical and biological sciences which, in the recent past, would hardly have been considered complementary.

The breadth of the computer program is indicated by the nature of the partnership which has grown up around it, a partnership which extends all the way across the United States. The Brain Research Institute is gratified to have established working arrangements with the Department of Engineering at UCLA and with the Center for Communication Sciences at the Massachusetts Institute of Technology where, under the leadership of our consultant, Professor Walter Rosenblith, much of the developmental work in this field has been done. A prominent member of the MIT team, Dr. Mary A. B. Brazier, has joined the Brain Research Institute staff, adding her own competence and leadership to the distinguished list of investigators already active in the program at UCLA. There can be little doubt that this study represents one of the true frontiers of modern science and, certainly, enlightenment regarding brain function will grow out of it.

Another Institute program contributes basic information to a second and related field which must be considered as a major modern frontier: the field of space science. Unfortunately, the national effort in space has been confined in the past almost exclusively to developments in the physical and engineering sciences, and little heed has been given to biology. All of that was changed when the Russian astronaut Titov displayed alarming reactions when aloft for only 24 hours. Now, as if for the first time, it is being realized that living systems are going to be subjected to the almost unbelievable stresses of a totally foreign environment in which acceleration, weightlessness, prolonged sensory distortion, disturbed biological cycles, isolation, radiation exposure and a host of other influences, acting independently and in summation, will critically challenge survival. Merely to provide for survival is not enough, for unless astronauts are to be considered merely drone passengers in the future as they have been considered in the past, they will be required to perform capably in space.

The Institute recognized a responsibility to examine the effects of space upon the chief organ of performance capability, the brain, by establishing a division called the Space Biology Laboratory. This Laboratory was founded and supported initially under contracts with and grants from the United States Air Force, but its programs now receive welcome additional assistance from the National Aeronautics and Space Administration, and from the National Institute of Neurological Diseases and Blindness. The Space Biology Laboratory, under the supervision of Dr. W. Ross Adey, devotes its attention to an examination of brain mechanisms when tested under conditions of simulated ballistic flight. The Laboratory has further devised studies utilizing animal subjects, which will permit a sophisticated assessment of brain performance in actual space invasion. The investigators are convinced that implementation of these studies will provide a precise and continuing method of determining the limits to which man's brain can be pushed in space without going to the extreme of utilizing man himself for the trials.

Another major Institute program is called the Clinical Neurophysiology Unit. In the past, few gulfs in science were broader than that which divided basic biology from clinical medicine. The National Institutes of Health recognized this fact when, in 1959, on a mandate from Congress, NIH established a program designed to provide facilities in which the techniques of basic research could be applied to problems of clinical practice. Such a program was established at UCLA under a grant from the National Institute of Neurological Diseases and Blindness. In this program the techniques of basic neurophysiology are utilized in studying patients with disorders that are susceptible to therapeutic intervention in the nervous system. As it happens, there is an increasing number of patients suffering from neurological and psychiatric conditions, such as Parkinson's disease, epilepsy, intractible pain and certain psychiatric disorders. Experience is proving that the neurophysiologist is a most valuable colleague in the clinical team assessing and treating these patients. In addition to increased knowledge of the basic disturbances underlying these disorders, these studies are also providing new insights into mind-brain relationships in man.

There are currently 86 projects and programs under way in the Brain Research Institute and I have described briefly only three. Obviously, it would not be feasible to review all of them. Perhaps more effective examples could have been selected from among other major interests of the Institute, such as neurophysiological correlates of behavior, molecular and cellular biology, radiation neurobiology, micro-and macromorphology of the nervous system, neuroendocrinology, psychopharmacology, neurochemistry, cerebrovascular mechanisms, marine biology, comparative neurophysiology, and many other topics including those particularly susceptible to specialized departmental techniques. Even though the research capability of the Institute is still gaining momentum, its vigor is indicated by the fact that well over one hundred publications a year emanate from its laboratories. Significantly.

greater contribution can be expected when recently occupied laboratories have become fully operational and when complete maturity of programs has developed.

A second principal goal recognized by the Brain Research Institute is the training of investigators for independent careers in research. Such training is one of the most important activities undertaken by the Institute, since available estimates indicate that research investigators in the biomedical sciences are in critically short supply. Whereas approximately 7,500 medical students are graduated each year in the United States, probably fewer than 500 students receive doctorate degrees in basic science related to medicine. Such an imbalance is most unhealthy and requires early correction if American research is to continue to lead the world in developing the health sciences.

The Brain Research Institute has committed itself to the task of training research investigators at both pre- and postdoctoral levels of development. Interest is centered prominently upon graduate students who are candidates for the doctorate degree, of which 30 now are receiving training in the Institute. Additionally, major activity has been established in postdoctoral research instruction of trainees of which 57 are now active in Institute laboratories. The magnitude of these training interests is indicated by the fact that nearly 200 investigators from 31 countries have participated in the training programs of the Institute during the past seven years. Some are senior scientists of distinguished accomplishment who contribute importantly to research undertaken in collaboration with members of the Institute. Others are students in an early stage of training who, after one to three years of experience, return home and establish new research activities in fields undeveloped in their own universities. All these visitors stimulate Institute programs and provide instructive information concerning foreign scientific developments. While these research and training dividends accruing from the foreign visitor program are substantial, humanistic rewards may be equally great: scientific associations of this kind provide a medium for exchanging ideas and encouraging understanding which avoids suspicions and antagonisms often implicated in social or political contacts.

The same combination of departmental and interdisciplinary effort which characterizes the research activities of the Institute is evident in the training programs. All graduate degrees are awarded by departments, but generous interdepartmental experience is provided for most trainees through preceptoral participation in collaborative research, as well as by means of presentation of a well-organized series of broadly interdisciplinary seminars and lectures.

Training activities of the Brain Research Institute center around a program entitled, "Interdisciplinary Program of Research Training in Basic Sciences Related to Mental Health," funded by a grant from the National Institute of Mental Health. This Program is just completing five years of operation, and a review of accomplishments during this period indicated it to be a highly effective instrument for scientist training. Ninety-six trainees (63 predoctoral, 33 postdoctoral) completed an average of two years training under the auspices of the program, and ninety percent of these individuals who have completed study on the program, have established academic and research affiliations in thirty institutions throughout the country.

The third principal goal of the Brain Research Institute relates to its activities in serving the social and scientific community through the development of many fields of brain study and through the dissemination of information regarding brain function. These activities take many forms, and it will be feasible here to review briefly only a few of them.

Initially, it is appropriate to mention the services which the Institute is attempting to provide its own investigators in order to encourage and assist their research and training activities. Co-ordination of financial, administrative and professional aspects of a large number of research programs is a responsibility of major proportions and one which the Institute approaches with the realization that solutions to many problems will mature only with time and with experience.

Prominent among services required by Institute programs must be mentioned the need for computer facilities, capable not only of processing existing requirements of investigators but also of developing new and refined methods of data processing. Computer equipment is costly and personnel, both technical and professional, is difficult to find. Fortunately, grant support from the Air Force and from the National Institute of Neurological Diseases and Blindness has provided a center for treating the special-purpose computer problems of members, while cooperative assistance from larger centers in the University and in the industrial community is supplying general-purpose computation needs of the Institute.

The rapid growth of the Institute and the development

of special requirements have created a situation in which other exacting needs of members have outstripped the ability of existing University facilities to provide certain services. The electronic requirements of Institute programs, for example, have made it necessary to equip special electronic shops for repair as well as for design and development. For comparable reasons the Institute has established mechanical shops where equipment can be designed and constructed. In addition, many investigators require facilities to process tissues, make chemical analyses, assist with the development of visual illustrations, type descriptive matter, edit manuscripts and provide other services which it is neither economical nor feasible for them to supply themselves. Such practical services as these, commonly underestimated by investigators and donors alike, are recognized by the Institute even though its ability to provide solutions to these problems is limited as yet.

A prominent commitment of the Institute relates to the development and dissemination of information regarding the brain by holding conferences and symposia at UCLA. At present, four such meetings are held annually around the general topics: The Brain and Behavior, Basic Research in Clinical Medicine, Computers and Communications Engineering in Brain Study, Brain Mechanisms and Space Sciences. Others are planned for the future. The proceedings of each of these conferences are published so that new information can be distributed widely on a continuing basis.

The Institute recognizes many other responsibilities. Public education, for example, is served by issuing appropriate press releases, by encouraging members to make general addresses and by periodically opening Brain Research Institute laboratories to conducted visits. Institute members serve on a multitude of local, national and international committees with assignments not only in science but in social development and in the humanities as well. They accept countless advisory commitments in the interest of local and national education, defense, international affairs and health programs. Further, as the Institute matures, it will find still other ways to contribute to the well-being of the community and society.

In the foregoing description I have attempted to provide a vignette so that many who have expressed an interest in the Brain Research Institute in the past, or will do so in the future, may understand something of its nature. New institutions commonly change the details of their dimensions as they develop, and it is recognized that the Brain Research Institute will focus its own image as time passes. But its goals will not change, nor will the devotion diminish which its members bring to the task of advancing these aims. Other periods and environmental circumstances have fostered the development of other institutes, but none of these has opened to greater opportunity than has the Brain Research Institute at UCLA.

WILLIAM F. WINDLE

Neuroanatomy in Relation to Experimental Neurology

IT IS APPROPRIATE in these opening exercises, dedicating the new Brain Research Institute of this University, that first consideration be given to neuroanatomy and its relation to experimental neurology. All of us realize that knowledge of form and structure is basic to understanding function. This is particularly true of the nervous system, which is quite complex. Techniques of anatomy cannot be separated from those of other aspects of biology that are brought to bear experimentally upon the nervous system. You will hear about neuroanatomy from other speakers today and tomorrow. It is difficult, indeed, to talk about any aspect of neurology without mentioning neuroanatomy.

Since I have been chosen to discuss this topic because of my identification for a good many years with both neuroanatomy and experimental neurology, it becomes my privilege to reminisce a little, and in so doing, to try to recall something of neuroanatomy in the recent past for a better appreciation of present trends.

Neuroanatomy had become a well-established subject

WILLIAM F. WINDLE is Chief of the Laboratory of Perinatal Physiology, National Institutes of Health, Bethesda, Maryland.

in the medical curriculum when I first became acquainted with it one day early in September, 1921. Professor Stephen Walter Ranson had provided me with a part-time job during my first year in medical school, and I became his research assistant and his pupil. The first experiment in which I played a small part had to do with vasodilator mechanisms involving effects of intra-arterial injection of ether in dogs—not very anatomical and only indirectly related to the nervous system. Soon, however, I was given an anatomical problem. Ranson had been unable to convince Professor Langley of Cambridge University of the existence of unmyelinated nerve fibers in the dorsal roots of spinal nerves. He had handsomely demonstrated their presence in peripheral nerves, spinal ganglia and spinal cord. No one, however, had been able to identify them in the roots themselves, a point which Langley emphasized in his denial of the afferent character of these small fibers. To him, they could only be sympathetics. It was gratifying to me to have had a hand in identifying these components of the peripheral nervous system. Ranson applied the anatomical knowledge of unmyelinated nerve fibers to physiological studies, especially in the conduction of pain and dissociation of pain from other sensory modalities. Since that time investigators have experimented along the same lines. Our present knowledge, nevertheless, is still incomplete, not because we lack the structural picture of nerves and their roots, but because we know neither what pain is nor how it is conducted.

As a pupil of Ranson, it was my privilege to experience the freshness of his approach to teaching neuroanatomy. He had just completed his famous textbook, which was the first to give the student a concise guide to the structure of the nervous system. I have re-examined my copy of the first edition and compared it with contemporary texts on the subject. I think the study of neuroanatomy forty years ago was much easier than it is today, not merely because there was less to study, but because Ranson presented the material with diagrammatic clarity and succinctness.

At that time, to be sure, there were some parts of the brain about which knowledge was scanty. The study of neuroanatomy involved memorizing numerous undefined terms, of which "cerebellum" was the least meaningful. One period of Ranson's course was devoted to the cerebellum. On that day we were delighted to hear that inadequacy of time to study more important regions demanded that we skip this dry chapter in the book. We did not learn much about the cerebrum either. Little was known; but for that matter, how much do we know about it today? (We hope that this new research center will quickly advance our knowledge of this part of the brain and of its role in human behavior.)

It is surprising that at the time when embryology was one of the few exciting fields of research in anatomy almost nothing was done with study of the nervous system. Embryologists were content to trace the formation of brain vesicles, to see major parts of the central nervous system unfold, and to follow the growth of peripheral nerves and sense organs. But they stopped short of exploring intrinsic development, and to this day we have only a fragmentary picture of the embryology of the human brain.

Advances in neuroanatomy have been made slowly. Neurophysiology forged far ahead in the 1930's. The cathode-ray oscilloscope, the electroencephalograph and other electronic devices made experimentation with the nervous system more appealing and, to many, more fun than studying its development and structure. Nearly every neuroanatomist tried to master neurophysiological techniques, and those who had not already become experimentalists did so. The morphologically oriented neuroanatomist almost became extinct. Only recently have neuromorphologists reappeared in numbers. These are the fine-structure neuroanatomists whose initial contributions have stirred the thoughts of every experimentalist. Electron microscopy, advances in neural histochemistry and revival of interest in tissue culture of nervous elements have ushered in a new epoch of neuroanatomy.

If there was a period during which neuroanatomists tried to become physiologists, or at least adopted the techniques of neurophysiologists, there is today evidence that the tables have turned. Neurophysiologists, pharmacologists, and even chemists are looking toward neuroanatomy to elucidate interrelationships of nervous and nonnervous elements, particularly at the submicroscopic level. The anatomist's inability to find among the cells of the brain adequate space to account for the extracellular space that the physiologist thinks he measures, has brought the two together. The revelation that nerve cells comprise only about one-fourth of all of the cells in the brain and that the more numerous neuroglia cells may be essential in the mechanism of active transport and have to be taken into account when metabolism is being investigated by the neurochemist, is another illustration of the new conjunction of fields.

Experimental neurology has been advanced by the development of other neuroanatomical techniques, such as those of Nauta and Rasmussen. Now that it is possible to follow the degenerating processes of nerve cells and now that we can render visible all synaptic endings on neurons, there has been a resurgence of experimentation on nervous pathways. The use of radioisotopes to follow migration of cells during developmental stages has helped renew interest in neuroembryology.

The methods of the neuroanatomist are no longer solely his. In this city there are psychiatrists performing exquisite experiments with one of the oldest of neuroanatomical techniques: that of Camillo Golgi. Also among my colleagues are surgeons, psychologists and chemists who are as facile with this and other methods as any neuroanatomist. Today if we ask who the neuroanatomist is, the answer will have to be that he is simply the one who belongs to the Association and teaches the subject, for his colleagues in other departments have assumed nearly all of his other attributes.

There has never been a time when so many people have been so actively engaged in seeking knowledge of the workings of the brain. This is reflected in the enormous increase in the number of publications in every form. Some look upon this trend with a jaundiced eye and deplore the effort required to keep informed about all the ripples in this vast sea of words. The contributions of the experimental neuroanatomist to this flood have thus far been relatively few, but their number is increasing. I predict it will not be long before he, too, will face the danger of submersion. It will take some skill to ride the crests; and the poor fellow who believes he must digest all neuroanatomical literature before qualifying as an investigator will surely drown.

The reason for this great scientific effort is known to all. There has never been a time when so much money has been so readily available to so many institutions to train investigators, support them in their careers, and pay for the increasingly costly materials and instruments now developed for their use.

To reminisce once more: Money for medical research at Northwestern University prior to 1926 was limited to about $12,000 per year. This represented the income from a single endowment, given in the hope of establishing a laboratory of experimental research in tuberculosis, but not used solely for that. The money was distributed among all departments of the medical school upon the request of individual staff members. Ranson received some, for his was one of the few active research programs. Other sources provided very little money for research in neuroanatomy. Grants-in-aid were almost unheard of and were very small by today's standards. As recently as 1945, the total annual institutional budget of Northwestern University's Institute of Neurology, a department devoted solely to research, was considerably less than is now required to purchase one electron microscope.

While it is evident that many outstanding contributions during the first quarter of this century were made by neuroanatomists with little more than an inquiring mind, a will to sacrifice and a few crude instruments, let us not be misled into picturing all neuroanatomists in this category. In a few places considerable sums of money were spent for research by neuroanatomists of the earlier years. For example, had Santiago Ramón y Cajal been unable to finance his research quite handsomely, he would not have made his great scientific contributions.

Prior to World War II, research in neuroanatomy was largely an individual, indeed a personal, affair. As an anatomist became established, he attracted graduate and medical students who served as his assistants, some of whom went on to neuroanatomical careers in teaching and research. Seldom did colleagues in other fields team up with the neuroanatomist. There was an especially wide gap between the preclinical and the clinical professors. Even after the revolution in investigative medicine was well under way, clinical research was thought to have little or no need for experimental animals. Therefore, no medical school building constructed prior to about 1945 provided adequate space for research laboratories in the clinical departments. It is more recently, and still in only a few places, that leaders in medical neurology and psychiatry have come to recognize the urgency of incorporating facilities for neuroanatomy, neurophysiology, neurochemistry and neuropharmacology into their own departments. This Brain Research Institute is a shining example of the conjugation of techniques in pursuit of knowledge of the nervous system.

Team research is quite new. It began to emerge in universities during the war, especially under such auspices as those of the Office of Scientific Research and Development. Since that time it has shown a steady growth not only in universities but also in governmental and industrial laboratories. There are many problems, among which are the really big problems of the nervous system, that cannot be solved by individual effort and require the cooperation of scientists from many fields.

In the balance of time I shall illustrate one collaborative effort by a team in our Laboratory of Perinatal Physiology. The problem concerns the relationships between adverse factors imposed during birth and the appearance of neurological disorders in the offspring. A number of people have participated; among them neuroanatomists, physiologists, biochemists, psychologists, pathologists, medical neurologists, pediatricians and obstetricians.

A number of human neurological disorders are thought to stem from adverse factors in prenatal, natal and early postnatal life. The more attention one gives to maternal-fetal relations, the more one is impressed with deficiency in knowledge concerning these periods. It would appear that ignorance of the physiology of the fetus, as well as of maternal gestational physiology, often breeds clinical mishaps that lead to mental retardation if not actually to neuromotor disorders such as those indicated by the term "cerebral palsy." The role of asphyxia in the etiology of neurological and psychological defects in the offspring has been debated for a century, and the literature is full of articles, reports and testimonials, pro and con. Even when one confines his attention to the few serious writings, one finds confusion in terminology and the lack of clear understanding of what takes place prior to, and during, birth. This is an area in which clinical research has been singularly ineffectual. Nevertheless, few animal experiments have been performed.

I shall limit my consideration to the role of asphyxia in the etiology of neurological defects in the offspring.

The principal experiments were conducted in monkeys (*Macaca mulatta*) of known gestational age near term. The entire contents of the uterus were removed during hysterotomy under local anesthesia. Generally the fetus was kept within its intact decidual membranes for periods of time varying from four and one-half to about sixteen minutes before being delivered.

Before describing some experiments demonstrating production of neurological disorders and before illustrating the neuropathology resulting from these experiments, I shall try to correct certain prevalent misconceptions regarding the physiology of the mammalian fetus.

A currently held misconception is that the human fetus normally develops in a milieu of low oxygen tension. This notion derives mainly from observations of the umbilical-cord blood obtained during birth, usually at cesarean section. Under these circumstances the fetal blood is poorly saturated with oxygen. The phrase "Mount Everest in utero" reflects this viewpoint. Blood passing from the placenta to the fetus is highly saturated with oxygen. Even after some admixture with venous blood in the fetal heart, the blood drawn from the carotid artery of the monkey fetus was found to be as much as 79 percent saturated. Blood returning to the placenta for reoxygenation was still about 58 percent saturated. These are values obtained by Dawes and his colleagues on blood samples drawn by catheterizing fetal vessels in utero. The fetal heart was beating at its normal rate, and glucose and lactate were not significantly altered until a state of hypoxia had developed, indicated by a decline in oxygen saturation to 40 percent or less. It would appear then that the monkey fetus is not short of oxygen and probably does not have to resort to anaerobic glycolysis under the conditions being considered.

Confusion exists concerning the use of the term "fetal apnea." Some believe that apnea at birth indicates that the fetus is hypoxic. This is not necessarily true; sometimes it is simply narcotized. As the result of studies on monkeys, as well as on other mammalian fetuses, we now recognize three types of fetal apnea: that existing normally in utero when the fetus is adequately supplied with oxygen from the placenta; that which during deteriorating conditions of oxygenation follows a transient interval of fetal arousal with shallow rhythmical respiratory movements; and the asphyxial apnea that follows the period of fetal gasping in the final stages of asphyxia. A fetus in asphyxial apnea will not begin to breathe spontaneously. It must be resuscitated without delay, and even though it may thereafter appear to be normal, one can expect it to have suffered brain damage.

Although our experiments fall into several categories, I shall limit this discussion to two: Firstly, animals which were asphyxiated until the final stage of apnea had been reached required resuscitation by intermittent admission of oxygen into the lungs under positive pressure through a tracheal cannula. Second, those which were asphyxiated for shorter periods of time and delivered from the decidual membranes just prior to cessation of fetal gasping did not require resuscitation, because the last respiratory efforts of the asphyxiated fetus admitted air to the lungs.

Monkeys asphyxiated and resuscitated at birth, exhibited a wide range of defective behavior, reflecting damage in the central nervous system. All showed motor weaknesses and disabilities during the first few days of life.

Certain marked defects persisted in varying degree in those animals requiring resuscitation. Almost uniformly the animals transiently lost the ability to suck and often to swallow. Feeding was a major problem and required round-the-clock nursing care. Abnormalities of locomotion, gait and balance were commonly encountered. The righting response was affected in nearly all. The young monkeys were hyporeactive and often lethargic. Although it is difficult to test for a lack of hearing ability, one monkey appeared to be totally deaf. Status epilepticus and persistent coma were encountered in a few. Ataxia, dyskinesia, paralysis, tremor, athetosis and other phenomena were found in severely affected monkeys.

Animals that were removed from their membranes during the stage of asphyxial gasping and required no resuscitation were weaker during the first few days of life than similarly delivered nonasphyxiated control monkeys. However, no serious neurological disorders persisted in these animals.

The brains of many of the infant monkeys subjected to asphyxiation and of several of the nonasphyxiated control specimens were studied histologically. The animals were killed at various ages by the perfusion-fixation technique, a procedure that assures instantaneous and equal preservation of all nervous tissue and that eliminates many artifacts seen in specimens fixed by immersion in formalin.

A consistent pattern of brain damage appeared in the monkeys that were asphyxiated at cesarean section and resuscitated after asphyxial apnea had set in. The lesion resulting from this procedure was a nonhemorrhagic, circumscribed focal cytolysis, present bilaterally and symmetrically. Small nerve cells were more vulnerable than large nerve cells and neuroglia cells. Capillaries remained intact. A microglial reaction set in within a few days and became marked by ten days. Ultimately the damaged region became a glial scar.

Approximately in order of severity, brain damage involved the following regions: nuclei of the inferior colliculi, sensory nuclei of the trigeminal nerve, ventral and lateral groups of thalamic nuclei, lateral cuneate and gracile nuclei, medial vestibular nucleus, portions of the putamen and globus pallidus and the subthalamic nuclei, and dentate and other cerebellar nuclei. Many other cell groups, including parts of the hippocampus, various central auditory nuclei, motor trigeminal and oculomotor nuclei, and the dorsal and intermediate cell columns of the spinal cord, were found to be affected.

In one animal, in coma for nine days before being killed, lesions were confluent and involved much of the lateral reticular formation of the midbrain and medulla oblongata, extending down into the sacral regions of the spinal cord. In most instances, however, the lesions were quite circumscribed and their boundaries sharply delineated from surrounding normal tissue.

Damage was limited to the nuclei of the inferior colliculi, thalami and sensory trigeminal nuclei, after very mild degrees of asphyxia. The brains of two animals, asphyxiated for six minutes or less, revealed no structural changes after two years. Those of three others, asphyxiated for seven to eight and one-half minutes but requiring no resuscitation, showed structural changes in the inferior colliculi and some other nuclei, but the changes were slight. Although structural brain damage was present, there had been no clinical signs of neuromotor disorders.

Only a few of the monkey brains had discrete lesions of the cerebral neocortex. The only animals in which such lesions were found had extensive damage in the thalamus and brain stem. These were animals that had suffered some setback in their health during the first few days after having been resuscitated successfully. It is possible that neocortical lesions may have been associated with some adverse factor in the early postnatal period, such as cerebral edema, rather than with the initial asphyxia. Similarly, few lesions of the cerebellar cortex were found and, when present, they involved the vermis.

The asphyxial lesions happened to be located in regions that in other species of animals have been shown to have a rapid rate of blood flow and presumably high oxygen utilization. But it is uncertain that they were produced solely by anoxia. The concomitant biochemical and hemodynamic changes during asphyxia may be equally or more important. To quote from a recent unpublished report by Dawes:

A limiting factor in asphyxial survival at birth might be the accumulation of acid metabolites, which would have the effect of reducing enzyme activity. It therefore seemed likely that if glucose and alkali (to neutralize the acidity) were infused during asphyxia, energy production might be increased and survival might be prolonged. Experiments on mature foetal monkeys showed that this is indeed so, and that the interval between the onset of asphyxia and the last gasp can be increased by more than twenty-five per cent on the average; recovery from asphyxia is also much more rapid after treatment.

Monkeys were given a standard fifteen-minute period of asphyxiation. Half received a slow infusion of sodium carbonate and glucose. The identity of those receiving and those not receiving infusions was unknown to investigators who examined them neurologically and those who later looked at the brains histologically. The animals receiving the infusion exhibited less marked neurological defects than those not receiving the infusion. The brains of all contained the characteristic lesions, but the extent and severity of damage varied in the expected direction. Thus, the way now seems open for a rational attack on one aspect of this problem. The neurological and histological defects resulting from asphyxia during birth are well enough established to permit a more intelligent approach to measures of prevention and therapy.

I have reviewed this one facet of experimental neurology to show how neuroanatomy may contribute. If I have not succeeded in making its role clear, it is because the techniques of neuroanatomy have become inseparable from those of other basic sciences. One may question whether the example chosen does not illustrate equally well research in obstetrics, respiratory physiology, neuropathology, or for that matter, pediatric neurology.

The project I have described represents only one of many efforts which are being made to explain the workings of the brain in health and disease, and to lay foundations on which to construct at some future time an understanding of the mind. It is fervently hoped that ways will be found to assure that the accelerating tide of scientific endeavor will continue to flow in this new Brain Research Institute and in others.

REFERENCES *

1. Bailey, C. J., and W. F. Windle, Neurological, psychological, and neurohistological defects following asphyxia neonatorum in the guinea pig, Exper. Neurol., 1:467–82, 1959.
2. Dawes, G. S., H. N. Jacobson, J. C. Mott, and H. J. Shelley, Some observations on foetal and new-born rhesus monkeys, J. Physiol., 152:271–98, 1960.
3. Dekaban, A., and W. F. Windle, Hemorrhagic lesions in acute birth injuries, J. Neuropathol. Exper. Neurol., (in press), 1962.
4. Hibbard, E., and W. F. Windle, Relation of birth asphyxia to cerebral hemorrhages, Physiologist, 4(3):47, 1961.
5. ——, Neurological consequences of a spontaneous breech delivery with head retention in a monkey, Anat. Record, 140:239, 1961.
6. Jacobson, H. N., and W. F. Windle, Observations on mating, gestation, birth and postnatal development of *Macaca mulatta,* Biol. Neonatorum, 2:105–20, 1960.
7. ——, Responses of foetal and new-born monkeys to asphyxiation, J. Physiol., 153:447–56, 1960.
8. Ranck, Jr., J. B., and W. F. Windle, Brain damage to the monkey, *Macaca mulatta,* by asphyxia neonatorum, Exper. Neurol., 1:130–54, 1959.
9. ——, Asphyxiation of adult rhesus monkeys, Exper. Neurol., 3:122–25, 1961.
10. Robert de Ramírez de Arellano, M. I., Maturational changes in the electroencephalogram of normal monkeys, Exper. Neurol., 3:209–24, 1961.
11. ——, D. L. McCroskey, J. M. Dennery, and W. F. Windle, Neurological deficits of asphyxia neonatorum in monkeys, Trans. Am. Neurol. Assoc., 84:151–54, 1959.
12. Shelley, H. J., Tissue carbohydrate reserves in foetal and new-born lambs and rhesus monkeys, J. Physiol., 151:24–25P, 1960.
13. ——, Blood sugars and tissue carbohydrates in foetal and infant lambs and rhesus monkeys, J. Physiol., 153:527–52, 1960.
14. Saxon, S. V., and C. G. Ponce, Behavioral defects in monkeys asphyxiated during birth, Exper. Neurol., 4:460–69, 1961.

* The references contain results of the principal experiments in the Laboratory of Perinatal Physiology, upon which the present article is based.

15. Windle, W. F., ed., *Neurological and Psychological Deficits of Asphyxia Neonatorum*. Springfield, Thomas, 1958.
16. ——, Effects of asphyxiation of fetus and new-born infant, Pediatrics, 26:565–69, 1960.
17. ——, Selective vulnerability in the central nervous system of rhesus monkeys to asphyxia during birth. Baden, Switzerland, CIOMS/NASA Symposium, (in press).
18. ——, H. N. Jacobson, M. I. Robert de Ramírez de Arellano, and C. M. Combs, Structural and functional sequelae of asphyxia neonatorum in monkeys (*Macaca mulatta*), Research Publ. Assoc. Research Nervous Mental Disease, 39:169–82, 1962.
19. ——, M. I. Robert de Ramírez de Arellano, M. Ramírez de Arellano, and E. Hibbard, Rôle de l'asphyxie pendant la naissance dans la genèse des troubles du jeune singe, Rev. Neurol., 105:142–52, 1961.

ALFRED E. FESSARD

Neurophysiological Frontiers in Brain Research

THE WORD "FRONTIER" has several meanings. It may mean the uncertain and fluid border separating knowledge from ignorance and, in that sense, the domain of brain research may be said to be in vigorous expansion nowadays. However, the pride of neurophysiologists should be somewhat attenuated when they consider the number of major problems that still remain unsolved. Among these, I would mention at random only three which are particularly resistant to the persevering efforts of the investigators: first, the origin of EEG waves, still obscure 35 years after their discovery despite a considerable amount of work; second, the nature of the synaptic transmitters for the basic processes of excitation and inhibition, for I would not commit myself to saying that acetylcholine is the only one; third, the problem of memory traces, i.e., of *where* and *how* they occur. This is the most fascinating but certainly the most difficult of all problems confronting brain physiologists.

However, a rich crop of new data has accumulated over the last twenty years. Some of it is of considerable

ALFRED E. FESSARD is Professor of Neurophysiology, Collège de France, Paris, France.

importance for our understanding of how the brain works and largely compensates for the temporary failures. The theme, *Frontiers in Brain Research,* offers an excellent opportunity for presenting some typical examples, among which I would choose a few in the field of neurophysiology.

Before doing so, I should not fail to call attention to another meaning of the word "frontier": the one designating the limits of any research field as defined by its object together with its technical aspects. Brain physiology is a research field that has much, but not all, in common with neurophysiology. On the one hand, there are other chapters of general physiology involved in brain studies: circulation, respiration, endocrine secretion, and so forth. On the other hand, the brain is only one part of the nervous system and it is not unimportant to ask whether all the basic processes discovered in the lower parts of the central nervous system are also those, and only those, underlying brain activity, or whether special properties other than the one of greater complexity would appear in the cerebral structures. Although the former assumption still prevails, it has sometimes been claimed that cortical neurons might be endowed with plastic properties not possessed by those in the spinal cord—an inference from observations of learning deficiencies in animals deprived of some or all of their cortex.

Whatever it may be, the frontier between general neurophysiology and cerebral physiology is wide open, and 'importations' of new data from the former domain to the latter are, as a rule, eagerly exploited by brain physiologists. We also know that the chart of spinal reflex action, established by Sherrington and now extended to processes at the neuronal level, is still a model for describing and analyzing the more complicated transactions within the higher nervous system.

The frontiers of brain physiology are also permeable to all that may usefully come from other disciplines, either basic or clinical. To limit ourselves to its relationships with basic sciences, I propose a diagram, shown in Figure 1, in which the seven most relevant fields are

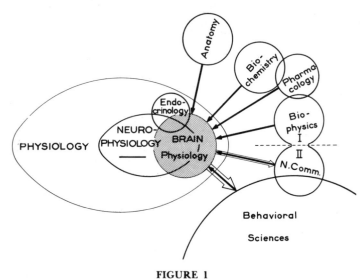

FIGURE 1
The most commonly recognized fields of basic research related to brain physiology.

represented. These are the disciplines selected by the 'International Brain Research Organization', an institution created mainly for the purpose of increasing and accelerating exchanges across the frontiers of the different research fields which serve the central one of brain physiology. Anatomy, biophysics, biochemistry and the broad domain of behavioral sciences are the principal disciplines, aside from general neurophysiology. Recent developments in neuropharmacology and neuroendocrinology have now justified their consideration apart from the others.

Nowhere does the artificiality of traditional frontiers among research fields appear so conspicuous as between brain physiology and brain anatomy, which in the past preceded and later inspired the first steps of the former. Today the alliance between neuroanatomists and neurophysiologists has become closer than ever, and such expressions as "anatomophysiology" and "electroanatomy" are indicative of this fact. On the other hand, neurophysiologists have greeted with great satisfaction and expectation the birth of neurochemistry, which sprouted not long ago from biochemistry.

The case of biophysics, which currently is in a phase of spectacular expansion, deserves special consideration.

The brain, like other organs, depends for its activity on intrinsic mechanisms that are described in terms of matter or energy exchanges. But some products of its activity belong to another 'dimension', that of *information,* a category that in recent years has received due recognition and given rise to special mathematical developments in the field of communication theory and engineering. An important subject has thus been made more explicit and rendered more accessible to scientific treatment. It is related to the distinction between *functioning* and *function;* in simpler terms, between the *how* and the *what for.* These two points of view are theoretically well recognized in general physiology, but in practice they are confounded in most investigations on living organs. The case of the brain is strikingly different.

The brain in its function of behavior control is essentially an organ that handles information for the purpose of communication and control. The term "neurocommunications" has been coined accordingly. The brain communicates with receptors on one side and with effectors on the other, and its integrative power is exerted by way of a multitude of intracerebral communications. Biophysics, in its relationship with the central nervous system, shows two faces. It goes as far as the present state of physics permits in the explanation of bioelectric potential—its generation and appearance—by introducing such processes as ion transport, impedance change, polarization, dipole, etc.; but it also attempts to apply the principles of the communication theory (in its symbolic expressions and its physical models) to a better understanding of the way in which the brain handles information.

More important than the conventional frontiers among the great disciplines at the service of brain physiology is, I believe, this formal distinction, this sort of demarcation line traced through the field of biophysics. Both sides of biophysics, labeled I and II in Figure 1, have formerly been put under the same heading simply because they require scientists of the same training in mathematics and physics. But these scientists, according to their orientation, are really dealing with two different universes of ideas and methods of data processing. Thus, in Figure 1 there are two circles to represent biophysics with an incomplete separation between the circles in order to indicate that operations performed in the upper sector—that is, measurements of physical phenomena—also involve in some sense the notion of information, because communication necessarily occurs between an observer and the facts revealed by the instruments used for measurements and recordings.

In this respect we are now facing the contrast between spectacular advances in the methods for automatic processing of electrophysiological data and the relatively slow progress in our applications of the principles and methods of communication engineering to the problems of information processing by the central nervous system itself. It has often been claimed that the brain and the computer are similar in their capacities for extracting and handling information, and in some sense this may be partly true; but a risk of confusion does exist here as concerns the kind and the amount of information with which either the brain or the machine have to deal. Information is an ambiguous term which must be related to the properties of the system that deciphers and utilizes the message. The discernible elements of a complex nervous activity cannot be recognized—and their probabilities of occurrence evaluated—in the same way by the decoding and integrating neuronal systems of the brain and by the electronic devices of the machine, primarily because what the computer is fed—a set of simultaneous EEG tape records for instance—is only an incomplete and distorted by-product of the intimate nervous activities from which the brain elaborates its patterns of action.

Such patterns command behavior, and animal behavior can be imitated by models, this being a task for biophysics II, as represented in Figure 1 by a zone overlapping the field of the behavioral sciences. Models are often built primarily for the purpose of giving a good approximation of some behavioral feature displayed by a living organism, and this may secondarily inspire some explanation at the level of neuronal mechanisms. However, there exists a reciprocal category of models whose internal constitution has been inspired by what is known of the most typical properties and interconnections of neurons and systems of neurons within the central nervous system. In both cases *information* is a useful concept, but in addition there is the involvement with the structure of the channels of distribution of the central nervous system. This theme may be said to belong to the domain of cybernetics. The two-way arrow in Figure 1 represents the reciprocal relationship between brain physiology and biophysics II.

For obvious reasons, a similar two-way relationship exists between brain research and the behavioral sciences. As a matter of fact, the bias is in favor of brain over behavior; the final task of brain physiologists is to contribute to an explanation of behavior in terms of underlying brain mechanisms. This certainly represents the most powerful trend of contemporary brain neurophysiology, especially in the United States, if we judge from the huge number of publications on the subject. But inferences from behavior to brain are also a method for gaining some insight into the cerebral mechanisms. That this method has to be taken as the only sound one for a study of higher nervous activity was, as is well known, claimed by Pavlov, who argued that a system whose essential function is integration and synthesis can only lose its specific properties when altered or mutilated by such methods as those of classical physiology.

One knows what a gigantic application of this principle has been made since Pavlov's time in his country, by the extensive use of the conditioned reflex technique. One also knows that this confinement to only one kind of approach, which has been the fate of our colleagues of the Soviet Union, was mainly due to the presence of another kind of frontier which prevented the exchange of ideas and personal contacts between scientists of the two worlds. That this frontier has become somewhat less impermeable in these last few years has been greeted with sincere enthusiasm, and has already resulted in unquestionable benefits to brain research.

The wisdom of the idea that a living brain is at every instant functioning as a whole is indisputable. It is certain that any experimental intervention in this most complex of all dynamic systems, even the administration of a light dose of an anesthetic, alters its properties and has often been the origin of false extrapolations as to its normal behavior. Electrophysiology has now, however, given us the possibility of recording panoramic aspects of cerebral activities in an awake and unrestrained animal. I venture to say that Pavlov, himself, would have acclaimed this technical achievement, which can go so far as to pick up activities from single cells during a conditioning or other behavioral situation, an achievement first performed on monkeys by Jasper, Ricci and Doane (12).

Such procedures do little or no harm to brain structures, and they represent a most remarkable analysis of some of the basic processes intervening here and there in the brain during a simple form of behavior.

However, we need more than that. Progress can proceed in two ways. Multiplying the number of implanted microelectrodes and using computers for data processing is one method whose future achievements we cannot foretell, especially when frequency modulation and phase shifts in repetitive unit activities from a number of different nuclei can be automatically correlated during behavioral situations involving higher nervous activity. But it is certain that this cannot go beyond certain limits, as millions of neurons are generally involved in simple acts and huge multiplication of electrodes would create destructive as well as unmanageable conditions. The other way is the continuation of our "acute" explorations despite the brain-as-a-whole credo, thus deliberately creating abnormal responses or behaviors. In this case, progress is to be expected not so much from an accumulation of new data as from a gradual improvement in our ways of reasoning when we try to cross this most treacherous of all frontiers: that from abnormal to normal and, what is worse, from treated dead brains to living brains, as when morphological distinctions are assumed to underlie functional features.

At this point we encounter another class of frontier that raises other acute problems. These are frontiers of the brain vis-à-vis the other parts of the central nervous system, and of course those within the brain itself, as revealed by the appearance of its heterogeneous components. Anatomists, histologists and cytologists led the way in this respect, and electrophysiology now inspires other partitions based on its own criteria. Furthermore, the time has come for biochemical and pharmacological procedures to propose their special cartographies (23).

Frontiers, thus created, essentially are relative to the technical operations that have engendered them and to the criteria retained for their recognition. Consequently, in general they do not coincide, and although they have in each case a descriptive value, they are not necessarily significant from the functional—in neurophysiology, the only essential—point of view.

This is not the place to recall in detail the various morphological subdivisions that have been proposed in the past, based first on gross appearance, then on direct or retrograde degeneration, myelogenesis, myelo- or cytoarchitectonics. The legitimacy, constancy and significance of the frontiers resulting from such subdivisions—cortical areas and transcortical layers, subcortical nuclear, laminar or reticular structures—are still a matter for passionate discussions. Since the pioneer work of Marshall, Woolsey and Bard in 1937 and 1941 (18, 19), and of Adrian (1) in 1941, neurophysiologists give more credit to electrophysiological criteria, especially to those based on evoked potentials. The primary projection areas are better defined as regions with initially surface-positive evoked potentials of shortest latencies than with reference to anatomical or cytological criteria. Associative territories display more labile evoked potentials of longer durations and latencies. With this powerful tool in their hands, associated with local stimulation procedures, neurophysiologists are now establishing new maps of the cerebral structures; they are not, however, protected from the risk of being misled in one way or another. For instance, barbiturate anesthesia restricts the cortical evoked potentials to those from the primary projection areas. This has for a long time remained an unrecognized fact, and the common use of this kind of anesthetic has long prevented neurophysiologists from being aware of the importance of responses taking place outside the specific territories, i.e., within the association areas. An anesthetic of opposite effect, chloralose, has revealed this important fact and, by its magnifying effect upon discrete normal activities, it has helped in the discovery, location and analysis of them in unrestrained chronic animals. This is the sort of problem to which my associates, Drs. Albe-Fessard (2, 3, 4), P. Buser (7, 8) and their respective collaborators have devoted most of their work in recent years (Figure 2).

Another common error is that of too readily attributing specific functional properties to one well-defined center, for instance, to speak of "the sleep center," when we know that such properties always belong to a system of interconnected nuclei. Perhaps as a counterpart to this way of thinking, another methodological trend has developed recently with the aim of making more apparent the system of interconnections which exists between different parts. Each center is represented as enclosed by a symbolic frontier and becomes a "black box," whose inner content is screened from our curiosity. Emphasis is now placed upon the set of connections, afferent and efferent, to which the center is assumed to owe one part of its functional properties, the other part resulting from the input-output transfer function of the black box. Organized systems of such boxes, the so-called "block diagrams," now blossom in neurophysiological publications. This is, should we say, for the best as well as for the worst.

A good knowledge of the communication pathways in the brain is unquestionably more essential than that of a huge number of separate territories whose frontiers are uncertain and subject to revision, as well as devoid of clear functional significance. In order to ascertain the role of these interconnections, the effects of their being cut through are often tested by neurophysiologists. Experimental frontiers are thus created by partial or total transsec-

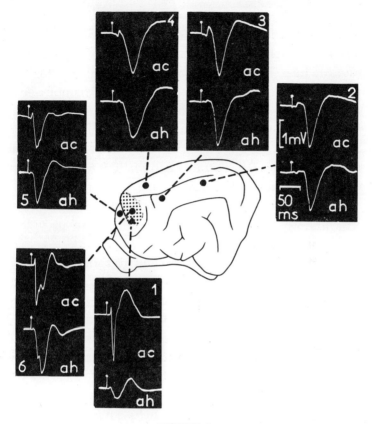

FIGURE 2
Cortical evoked potentials from different regions of a cat's brain (chloralose anesthesia), as shown on the central schema (the dotted zone is the primary somatic area). The stimulation is either that of the contralateral (ac) or homolateral (ah) foreleg. Record 1 is that of the primary evoked potential; the other records correspond to the five optimal zones for nonprimary (bilateral) projections.

tions through connecting structures. The best example—and this concerns the most fruitful recent studies in this respect—is that of the brain stem. Spinal cord (on one side), brain and chiefly cerebral cortex (on the other), had for long been the preferred objects of central neurophysiologists. The brain stem between the spinal cord and the brain appeared to be little more than a mere *voie de passage* for an intense traffic both ways, along private communication lines. Had this been the case, transsection through the brain stem should have resulted in deficits not differing essentially, though differing quantitatively, at various levels of transsections. Instead, we now know what dramatic effects are observed, for which location of the section is very critical. "Decerebrate rigidity," for the lower part of the body, and cortical sleep, in the Bremer *cerveau isolé* preparation, are the well-known spectacular symptoms for which the old classical notions have failed to provide a satisfactory explanation.

With the discovery by Magoun and his collaborators (16, 17) fifteen years ago of the antagonistic roles of the pontomesencephalic and of the bulbar reticular formation, respectively exerting descending facilitatory and inhibitory actions, the first symptom received its long-awaited explanation. Three years later the second symptom was no less clarified when Moruzzi and Magoun (20) disclosed the arousal effect produced by stimulation of an ascending activating system which was also included in the reticular formation of the brain stem. A new era of fruitful investigations, in search of other critical frontiers within the brain stem, opened with these epoch-making discoveries. In Figure 3, some characteristic levels of transsection are compiled.

Frontiers endowed with functional significance have thus been disclosed at different levels of the reticular formation. Apart from the wide ones enclosing the facilitatory and inhibitory zones, it appears that the former zone contains specialized nuclei (see Olszewski, 21) of which two at

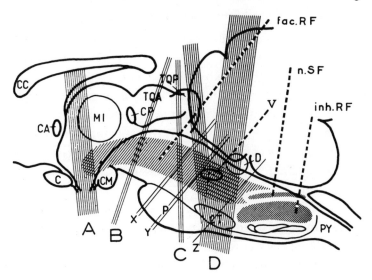

FIGURE 3

A schema of the brain stem bringing together different levels of transsection as performed in neurophysiological experiments. TQA and TQP, superior and inferior colliculi; fac. RF and inh. RF, facilitatory and inhibitory portions of the reticular formation; n.SF, nucleus of solitary fasciculus; CT, trapezoid body; P, pons; CM, mammillary bodies; V, chief sensory nucleus of the fifth nerve; B, classical transcollicular section of the Bremer "cerveau isolé" preparation; A and D, critical levels for the control by adrenaline (injected intravenously) of the monosynaptic reflex (facilitation after section at any of the levels at A, inhibition after section at any of the levels at D); C, the most rostral sections for ascending activating effects of adrenaline injections (9); X and Y, limits of transsection for the observation of enduring EEG activation patterns (5); Z, retropontine section just separating the n. *pontis centralis caudalis* from lower levels, and after which all peripheral signs of rhombencephalic sleep disappear (14). (Adapted from Dell, Bonvallet and Hugelin, 9)

least have been given recently a physiological role: the *nucleus pontis centralis caudalis*, which Jouvet (14) identified as the main part of the rhombo-rhinencephalic complex involved in the so-called "paradoxical phase of sleep" he described three years ago (13); and the *nuclei gigantocellularis*, which Bowsher and Albe-Fessard (6) have shown to be one of the main relay stations for the paleospino-thalamic tract, from which nociceptive impulses are projected to the parafascicular-centrum medianum complex of the thalamus in the cat's brain. Block diagrams are easily and usefully drawn in such cases.

In the thalamus, composed of closely packed nuclei whose frontiers are not always well defined by the faint outlines that appear on the cytoarchitectonic microphotographs, it is clear that the block-diagram representations are fairly inadequate; and the technical problem is no longer one of mapping out plans of interconnections but one of exploring the content of the "black boxes" themselves. In other words, the technical approach must be that of sampling the neuronal populations with microelectrode recordings. Fortunately, the brain seems to work statistically in the lowest level of its subdivisions so that we may, at least to a certain extent, extrapolate from samples to the properties of the whole population.

Stereotaxic maps prepared from cytoarchitectonic microphotographs are the necessary guides for such explorations. However, the real task of neurophysiologists is not that of checking the frontiers proposed by the histologists. It is mainly the task of trying to correlate these purely morphological separations with corresponding discontinuities in local dynamic properties (potentials, threshold, latencies, etc.) and in some behavioral aspects, insofar as the explored structures take part in the neurophysiological antecedents of an overt reaction. Actually, the normal steps in a complete neurophysiological approach to some particular operation performed by the brain should start at neuronal levels and end with behavioral manifestations.

The following example taken from a series of recent investigations carried out in Dr. Albe-Fessard's laboratory with the help of some collaborators illustrates this principle. In this instance, the frontier is between two thalamic nuclei: the ventrobasal (VB) complex, a somatotropic relay to the cortex; and the dorsally adjacent field of neurons generally attributed to the n. ventralis lateralis (VL), preferably called VL complex. A Nissl staining photomicrograph of a coronal section would show the outlines of the VB complex; whereas, the VL complex would appear as a region of lesser cell density. That the frontier between these two nuclei is also a frontier for evoked potentials is shown in Figure 4 (15).

These differences can be explained on the basis of a duality in the somatic afferent pathways: afferents that correspond to the short-latency, somatotropic responses consist of the dorsal columns of the spinal cord and the medial lemniscus; afferents that are exclusively of the anterolateral tract, the so-called "spinothalamic pathway," are relayed by extra-lemniscal routes. The bilaterality of the responses in the VL complex is consistent with this explanation, as is also the fact that purely tactile messages, and those from the receptors in the joints, do not project to the VL complex. They are, however, relayed in the VB complex. By contrast, the part of the VL complex we are concerned with is activated when pin pricks or sharp blows are applied to the skin, and also when other sorts of nociceptive stimuli are used.

In order to see if there exists a behavioral correspondence to the marked electrophysiological as well as structural differentiation between the two nuclear complexes, VB and VL, chronic animals with implanted electrodes were prepared. An electrode was placed just at the limit between these nuclei. This was technically possible because

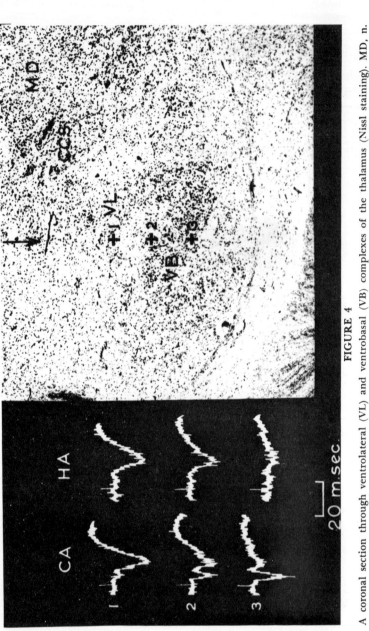

FIGURE 4

A coronal section through ventrolateral (VL) and ventrobasal (VB) complexes of the thalamus (Nissl staining). MD, n. medialis dorsalis; CCS, central commissural system. On the left, tracings correspond to three successive placements of the recording monopolar electrode. The arrow indicates the direction of the electrode track, not clearly visible here. On the left are shown three pairs of evoked potentials picked up at 1 mm. distance. They result from stimulation of the skin of the contralateral (CA) and homolateral forelimbs (HA). Note the transition of the long-latency, bilaterally evoked VL response to the short-latency, contralateral VB response. Striking differences in latencies, durations and shapes are thus observed, with the intermediate contralateral response an obvious combination of the two extreme ones. (Reproduced from

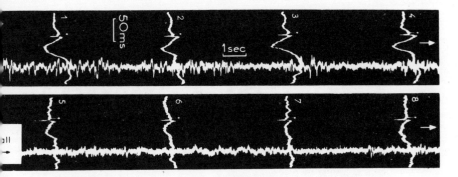

FIGURE 5

Unrestrained waking cat with permanent electrodes. Horizontal trace: spontaneous activity of the cerebral suprasylvian gyrus. A "call" is followed by the ordinary arousal reaction. Vertical traces: intermittent evoked potentials at higher sweep speed recorded at the border of VL and VB complexes, so that a composite record (brief positive deflection plus slow wave) is obtained. Observe the difference in the reaction of the two components when the cat is aroused.

of a device permitting fine adjustments of the electrode depth after it had been approximately placed under stereotaxic conditions and permanently fixed to the skull. The control for good placement was the occurrence of a two-component evoked potential. A small branch of the radial nerve was regularly stimulated by way of permanent electrodes buried in the tissue. The stimulation was apparently nonpainful and, in fact, was disregarded by the cat. Figure 5 shows how differently the two structures react when the animal, initially in a drowsy state, was aroused by being called.

This is also supported by another result obtained by applying a simple conditioning procedure (sound plus nonpainful electric shock) while recording either from the VB or from the VL complex in a chronic animal. This again reveals the big difference that exists between these two nuclei with respect to the way in which they take part

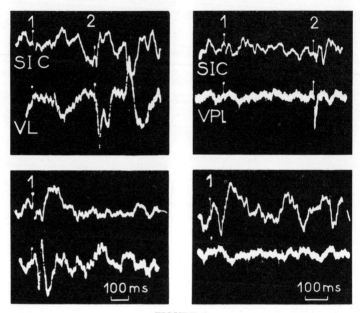

FIGURE 6

Double trace recordings during conditioning experiments on the same cat. Upper trace (SIC): activity of the primary somatic area in a zone of convergence. Lower trace: left side, activity in the VL complex; right side, activity in the VB complex (VPL). Conditioning is obtained by pairing a sound (1) and an electric shock (2). In each column, the double records above correspond to the initial pairing, the double records below to a moment when the link has been established (in VL) or has proved to be impossible (in VPL). (From unpublished results by G. Lelord and D. Albe-Fessard)

in a functional activity of a more complex order. In these experiments it has never been possible to condition the VB complex, while the VL nucleus of the same animal reacts to sound alone after the appropriate number (between 15 and 35) of paired stimuli have been applied (Figure 6).

Neurophysiology now extends its field of investigations to the microscopic level. In a case like the one previously presented it is not unreasonable to raise the question: How is the neuronal population organized at and near the frontier between the two neighboring nuclei? *A priori* representations may endow the constituting neurons with sharply different properties and may picture them as partly interspersed in a mosaic-like zone corresponding to the place where composite tracings are recorded (see Figure 3). There may, in addition, be neurons endowed with mixed properties variously distributed on each side of the border. To get an idea of how this actually occurs, an exploration at unit level was undertaken. For each unit tested with a microelectrode, the receptive field was ascertained and the nature of its adequate stimulus, either purely tactile or otherwise, was determined. It could thus be shown that three categories of units exist: a) the purely specific or primary ones; b) those toward which only extra-lemniscal afferents converge; and c) mixed ones.

The distribution of a sample of such units is shown in Figure 7. As can be seen, no unit within the primary relay receives purely extra-lemniscal afferents, but a certain proportion are of the mixed category. The central core contains only purely specific ones. On the contrary, beyond the frontier the extra-lemniscal category predominates, and most of the units respond with an early plus a late discharge (double response). Figure 8 is an example of a mixed unit in VPL which reacts electively to the bending of a definite joint but can also respond—here by an inhibitory reaction—to a tap on a localized part of the skin.

Similar investigations have been extended to many other territories possessing multisensory receptive fields, where each neuron is characterized by its proper set of afferences. They do not necessarily respond by delivering spikes, but often only by subthreshold excitatory synaptic potentials, or inhibitory ones, which simply modulate the rate of a

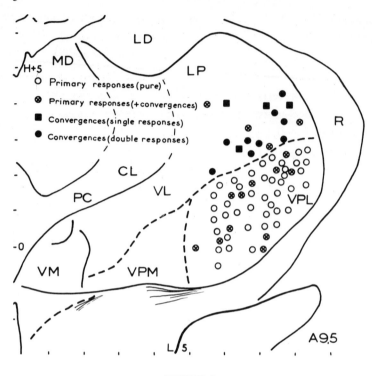

FIGURE 7
A sample of units in VPL and VL responding to different categories of afferent impulses. Light chloralose anesthesia. (From unpublished results of Mallart, Albe-Fessard and Martinoya)

permanent autogenic rhythm. Chloralose helps to identify these neurons and their characteristic afferent patterns by apparently lowering their firing level. It obviously, however, does not create new connections. This multivalence of so many neurons, as found in recent years by an extensive exploration of cortical and subcortical structures, was greeted by some neurophysiologists as a very promising revelation. It undoubtedly has something to do with the phenomena of information-processing in the central nerv-

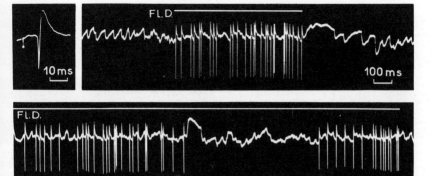

FIGURE 8

Activity of a mixed unit in VPL responding to the flexion of a particular claw (FL.D., upper tracing). The continuous discharge once this claw is bent (FL.D., lower tracing) is interrupted when sharp pressure (Pr.P.) is applied locally to the foreleg. Left upper corner: fast sweep record of a response to an electric shock.

ous system, especially when integrative operations are performed. Heterosensory integration seems to be the necessary underlying operation at neuronal levels for the formation of new nervous associations during learning procedures. The notion of "frontier" reappears here in the sense that new synaptic gates are crossed while others are closed (10, 11).

What by far constitutes the most basically important frontier in the central nervous system, that of the nerve cell itself, is thus evoked and could give way to interesting discussions. Molecular and ionic transits through this well-protected and specifically organized border are the fundamental processes of all nervous transactions. However, we must consider this topic one for general neurophysiology and biophysics rather than for brain research. Let us recall only that in addition to the natural ways (chemical or electrical) by which the neuronal membrane is made func-

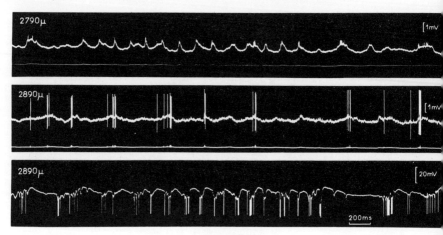

FIGURE 9

Spontaneous activity of the same cortical unit in the somatosensory area of a cat's brain at two levels of penetration. The two upper records each consist of two simultaneous extracellular tracings, of which the upper tracing is at high amplification, and the lower tracing is at low amplification. The third (bottom) record is intracellular and at low amplification. It shows the sign reversal of both spikes and slow waves and the predominance of spikes over slow waves; this is in contrast with the predominance of slow waves in the upper record, at the level of the dendritic tree.

tionally permeable at some specific loci, physiologists have become skillful enough to be able to violate nerve cell frontiers by using very fine puncturing microelectrodes that do not destroy the cell immediately. This procedure applied to neurons in situ within suitable cerebral structures, i.e., those cells sufficiently large, is an important source of information, mainly concerning the sign, true shape and origin (external or internal to the impaled neuron) of slow potentials. An example in which autogenic slow waves can be assumed to originate in the dendritic tree of an impaled pyramidal neuron is shown in Figure 9. It is a case where the traumatizing effect of the puncture does not seem to have significantly altered the pattern of discharge, but this unavoidable lesion to the membrane often has consequences to be taken into serious consideration.

I would end this series of general remarks centered around the notion of "frontier" by recalling that the organizing factors of the functional activity of a working brain are basically in the form of patterns appearing in an ever-changing distribution of a multitude of elements, a representation that leads to the fading of all that may deserve the name, "frontier," if we except that of the neuron itself. The dynamic properties of neuronal populations implicated in the activity of any part of the brain can be described by three superimposed categories of fluctuating patterns: that of their active "trigger zones" at every instant; that of the actual synaptic afferents on the somatodendritic surfaces of each neuron; and that formed by the group of neurons fed by the branches of each single axon. The second pattern is the determinant for the configuration of the first, which in turn determines the distributions involved in the third.

Behind each kind of particular behavior or mental representation there must exist a multiplicity of such patterns, well-adjusted or congruent to one another. Changes in them from instant to instant occur in an orderly way. The whole tune is played by millions of travelling impulses, themselves ordered in time patterns, whose role is to force or close multitudes of synaptic gates. All this has been expressed more than once, and is once more evocative of the familiar lock and key analogy, unless one prefers the more modern and more elaborate one of punch cards. To quote Sherrington (22): "The question who turns the key, to use that metaphor, is soon answered; the outside world." The brain neurophysiologist, after he has well recognized the frontiers of his domain, has essentially to deal with patterns. His task is analogous to that of a locksmith who wants to discover the code of a combination lock he has not made. This applies to the neurophysiologist who stands half-way between the biophysicist and the psychologist, but we know that all grades exist between these two extremes and that neurophysiologists today constitute a very heterogeneous family. No insuperable frontiers exist, however, between its members, who are united for the immense task and fascinating adventure of exploring and explaining the brain.

REFERENCES

1. Adrian, E. D., Afferent discharges to the cerebral cortex from peripheral sense organs, J. Physiol., 100:159–91, 1941.
2. Albe-Fessard, D., and A. Fessard, Intégrations thalamiques et leurs conséquences au niveau télencéphalique, in *International Colloquium on Specific and Unspecific Mechanisms of Sensory-Motor Integration.* Pisa, 1962.
3. ——, and A. Mallart, Existence de réponses d'origines visuelle et auditive dans le centre médian du thalamus du chat anesthésié au chloralose, C. R. Acad. Sci., 251:1040–42, 1960.
4. ——, and A. Rougeul, Activités d'origine somesthésique évoquées sur le cortex non-spécifique du chat anesthésié au chloralose. Rôle du centre médian du thalamus, EEG Clin. Neurophysiol., 10:131–52, 1958.
5. Batini, C., G. Moruzzi, M. Palestini, G. F. Rossi and Zanchetti, Effects of complete pontine transsections on the sleep-wakefulness rhythm: the midpontine pretrigeminal preparation, Arch. Ital. Biol., 97:1–12, 1959.
6. Bowsher, D., and D. Albe-Fessard, Patterns of somatosensory organisation within the central nervous system, (in press), 1962.
7. Buser, P., Observations sur l'organisation fonctionnelle du cortex moteur chez le chat, Bull. Acad. Suisse des Sciences médicales, 16:335–97, 1960.

8. ——, P. Borenstein, and J. Bruner, Étude des systèmes "associatifs" visuels et auditifs chez le chat anesthésié au chloralose, EEG Clin. Neurophysiol., 11:305–24, 1959.
9. Dell, P., M. Bonvallet, and A. Hugelin, Tonus sympathique, adrénaline et contrôle réticulaire de la motricité spinale, EEG Clin. Neurophysiol., 6:599–618, 1954.
10. Fessard, A., Le conditionnement considéré à l'échelle du neurone, Colloquium on Electroencephalography of Higher Nervous Activity (Moscow, 1958), EEG Clin. Neurophysiol., (Suppl.) 13:157–84, 1960.
11. ——, and Th. Szabo, La facilitation de post-activation comme facteur de plasticité dans l'établissement des liaisons temporaires, in Delafresnaye, J., ed., *Brain Mechanisms and Learning*, CIOMS Symposium, pp. 353–73. Oxford, Blackwell, 1961.
12. Jasper, H., G. F. Ricci, and B. Doane, Patterns of cortical neuronal discharge during conditioned responses, in Wolstenholme, G. E. W., and C. M. O'Connor, eds., *Neurological Basis of Behavior*, pp. 277–94. London, CIBA Foundation Symposium, 1958.
13. Jouvet, M., and F. Michel, Recherches sur l'activité électrique cérébrale au cours du sommeil, C. R. Soc. Biol., 152:1167–70, 1958.
14. ——, Mise en évidence d'un centre hypnique au niveau du rhombencéphale chez le chat, C. R. Acad. Sci., 251:1188–90, 1960.
15. Kruger, L., and D. Albe-Fessard, Distribution of responses to somatic afferent stimuli in the diencephalon of the cat under chloralose anesthesia, Exp. Neurol., 2:442–67, 1960.
16. Magoun, H. W., Bulbar inhibition and facilitation of motor activity, Science, 100:549–50, 1954.
17. ——, and R. Rhines, An inhibitory mechanism in the bulbar reticular formation, J. Neurophysiol., 9:165–71, 1946.
18. Marshall, W. H., C. N. Woolsey, and P. Bard, Cortical representation of tactile sensibility as indicated by cortical potentials, Science, 85:388–90, 1937.
19. ——, Observations on cortical somatic sensory mechanisms of cat and monkey, J. Neurophysiol., 4:1–24, 1941.
20. Moruzzi, G., and H. W. Magoun, Brain stem reticular formation and activation of the EEG, EEG Clin. Neurophysiol., 1:455–73, 1949.

21. Olszewski, J., The cytoarchitecture of the human reticular formation, in Delafresnaye, J. ed., *Brain Mechanisms and Consciousness*, CIOMS Symposium, pp. 54–80. Oxford, Blackwell, 1954.
22. Sherrington, C., *The Brain and its Mechanism*, p. 7. Cambridge Univ. Press, Cambridge, 1933.
23. Vogt, M., The concentration of sympathin in different parts of the central nervous system under normal conditions, J. Physiol., 123:451–81, 1954.

SEYMOUR S. KETY, M.D.

Regional Neurochemistry and Its Application to Brain Function

WHAT IS OFTEN REGARDED as the young science of neurochemistry is not as young as one might think. It is, however, a science which has had a long adolescence and a long period of bachelorhood, and which has only recently married and produced offspring. As a matter of fact, Thudichum, who is regarded as the father of modern neurochemistry, received from the British government about one hundred years ago a grant to pursue a research program on the chemical composition of the brain. Actually, neurochemistry has had a thin and sparse history from that time until quite recently. Its early development was colored by the problem of winning independence from its maternal science, i.e., of separating neurochemistry from biochemistry and of justifying its independent existence by differentiating the brain as an organ from the other organs of the body. Thudichum (47) concentrated on the brain as a whole in studying its composition, although he did recognize and sometimes characterize white matter and gray matter separately. A little later Mosso, the physiologist, advanced neurochemistry beyond composition. He inserted

SEYMOUR S. KETY, M.D., is Director of the Department of Psychiatry, Johns Hopkins University, Baltimore, Maryland.

very sensitive thermometers into the brains of living animals, and concluded from his measurement of brain temperature that the brain was the focal point of a very active metabolism (36). After the work of Barcroft and Warburg, brain slices, minces and homogenates increasingly became objects of study, but these specimens were concerned with study of the whole brain or at best with that of the cortex and white matter (13). Following that, there developed an interest in studies made in vivo which attempted to get at the metabolism of the brain in the living animal or in man (23, 27, 28, 41).

This preoccupation with the brain as a whole did not stem from a lack of appreciation of its heterogeneity, but rather was derived from a recognition that there was much about this organ and its metabolism which distinguished it from other organs of the body. Much, indeed, was learned in this adolescence of neurochemistry. We learned about the existence of the blood brain barrier (3), a phenomenon which is rather unique for the brain, and about the control of the cerebral circulation (45), which is quite different from circulation elsewhere. In confirmation of Mosso's prediction, it was learned that the brain is the focal point of one of the most active metabolic engines in the body and, as opposed to practically every other organ in the body, has a respiratory quotient of unity. This suggested what was later confirmed: that glucose is the substrate for the brain's never ceasing energy requirements and that without glucose the brain could not function normally (28). This again marked the brain from other organs. For example, the heart is not at all dependent upon glucose, for it can utilize, and in fact does utilize, fatty acids or lactate and pyruvate from the blood in preference to glucose (19). The brain, at least in the living state, is unable to do so.

Even these major distinctions, however, were not readily nor widely appreciated. Biochemists were reluctant to recognize differentiation and specialization, preoccupied as they were at that time with the so-called "unity of biochemistry." This concept or doctrine that biochemistry was fundamentally the same in every living cell permitted a certain faith in the ultimate simplicity of the animal organism but unfortunately cultivated a certain blindness to the potential wealth and variety in biochemistry.

Meanwhile, some neurochemists from other disciplines, who had been raised on the complexity of the brain, were beginning to make localized observations within the brain. Serota and Gerard repeated the early observations of Mosso, but this time with very careful regard for the localization of the temperature measurements, and found in the brain significant temperature gradients which they felt could be related to functional activity (42). Histochemistry developed very early from histology and perhaps from an attempt at a chemical understanding of histological staining reactions. Nearly two decades ago, Lowry and Pope separately began to map the brain's uttermost detail with precise enzymatic measurements, highly specialized with reference to particular regions. Pope applied the techniques of Linderström-Lang and studied conglomerates of cells in the brain with reference to their anatomical position on the one hand, and with reference to certain important enzymes on the other (37). Lowry carried this kind of procedure to a more exquisite level of resolution (34). He developed techniques for dissecting out, weighing and analyzing individual neurons or parts of neurons for the presence of a large variety of enzymes and more recently of substrates.

After the exploitation of the brain as a whole, neurochemistry is now ready to recognize and to investigate the most important way in which the brain is unique: its magnificent organization. The great advances in biochemistry were made to a very large extent upon two organs, muscle and liver, in which processes within the cell are an excellent representation of the functions of the organ itself. A liver slice or a muscle spindle go on being liver and muscle in the Warburg flask or in the perfusion chamber and, at the same time, they permit the study of the biochemical processes which underly their normal functional activity. But the brain, more than anything else in the universe perhaps, is far greater than the sum of its parts. Most of its special function stems not only from its component cells and cell masses but also from the connections and interrelationships among them. The brain slice has lost its most essential ingredient—its connection with the rest of the brain, with the body and with the outside world.

I should like very hastily, and without any pretense of being exhaustive or even fair, to indicate some of the activity which is going on in attempts to learn more about the regional biochemistry of the brain and to relate these findings to function.

We may turn first to regional metabolism. Of course, following the studies upon the brain slice, others began to dissect the brain and to study the respiration of the different parts of the brain in the Warburg flask. Some years ago my laboratory became interested in this question, recognizing that perhaps the optimum approach to metabolism, even at the level of energy transfer of the different parts of the brain, would eventually have to be the living brain.

As a first approach to an understanding of local oxygen consumption, we felt that it was necessary to develop techniques for the measurement of regional circulation under conditions of physiological normalcy. In collaboration with Sokoloff, Landau, Freygang and Rowland (32, 37), we were able to work out a technique which permitted semiquantitative measurement of the perfusion rates of various parts of an animal brain of rather small dimensions during physiological activity. The measurements themselves were made after the animal was sacrificed. In fact, one of the difficulties of the original technique was that we had to kill the animal in order to make a group of measurements, a procedure which made serial measurements difficult. But although the determinations themselves were made upon brain sections, they related to a period of about one minute before the sacrifice of the animal, during which the animal could have been, and hopefully was, maintained in a reasonable state of normality. This technique depends upon the uptake of a radioactive diffusible substance by the brain. It is quite apparent that if a diffusible substance is made available to all the tissues in the body by its presence in arterial blood, the quantity of that substance taken up—or the concentration achieved in any small homogeneous area—would be a function, among other things, of the circulatory rate through that structure. By measuring other variables or by maintaining them constant, it was possible to relate the concentration of the radioactive diffusible substance in various parts of the brain to regional cerebral blood flow. Figure 1 is an autoradiogram of a section of the brain of a cat that was exposed to this radioactive gas in an unanesthetized state. The density is produced by the radioactivity in each area, and it can be shown that it is roughly proportional to the blood flow there. One can see how well the cortex stands

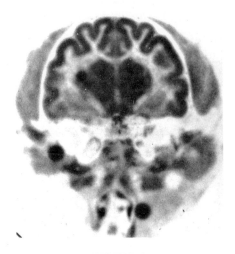

FIGURE 1
Autoradiogram of a frozen section of the brain of a cat after exposure of the animal to a radioactive gas. (See text for interpretation.)

out from the white matter, and how the thalamic nuclei can be differentiated in a vague sort of way—especially if one is not an anatomist. Interestingly enough, the visual cortex stands out more markedly than do other areas of the cortex.

Figure 2 is a more caudal section through the occipital lobe and portions of the brain stem. The finding of great interest in this figure is the inferior colliculus, which characteristically in these autoradiograms stands out more than any other structure in the brain, along with the superior olive and a collection of nuclei, perhaps the nuclei of the lemniscus which lie between them.

Dr. Windle has mentioned his finding of damage in the inferior colliculus of asphyxiated neonatal monkeys. This damage may be related to the very high perfusion rate in the inferior colliculus, a condition which suggests a high

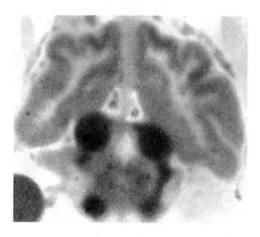

FIGURE 2
Autoradiogram of a more caudal section than Figure 1. (See text for interpretation.)

metabolic rate. To the best of my knowledge this particular finding was not anticipated by any other work, and we were surprised to find that the inferior colliculus had the highest circulation in the brain, representing a blood flow about three times the value for the brain as a whole and nearly twice as great as that for much of the cortex. We have no explanation for this high rate of flow. At first we thought that it might have been so high because the cats were not anesthetized and because their auditory systems were responding to the frequent clicks associated with the various counters we were using to monitor the radioactivity. For this reason, Landau destroyed the middle ear of a small series of cats, and after making quite sure that the animals were deaf, he still found that the inferior colliculus had an extremely high blood flow. The importance of these observations is not in terms of blood flow alone. Because there is considerable evidence that blood flow in the brain is usually well adjusted to metabolism, these measurements give an index of that function. In a way these slides have represented a metabolic map of the living brain, indicating the differentiations in terms of oxygen or energy utilization for the various parts.

Figure 3 is a summary of the effects of anesthesia upon various regions of the cerebral cortex. Notice that in the unanesthetized state the somatosensory, the auditory and the visual cortex have a rather rapid blood flow, suggesting a high rate of metabolism as opposed to the olfactory and miscellaneous cortex. This, we think, is indicative of a greater degree of functional activity in association with the conscious state. Interestingly enough, there is a slight fall in the blood flow and presumably in the metabolism in all areas of the brain under anesthesia, but a much greater fall is in the most active areas of the cortex, so that the contrast between the primary sensory areas and other regions of the cortex disappears in that state.

Sokoloff (44) has shown that these patterns of blood flow and presumably of metabolism are quite sensitive to changes in functional activity. By exposing some of the cats to photic stimulation for a period of five minutes before the measurement was made, it was possible to highlight the nuclei in the optic pathways. The lateral genicu-

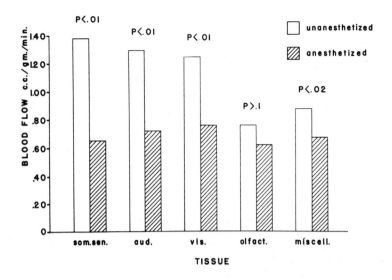

FIGURE 3
The effects of thiopental anesthesia upon various regions of the cerebral cortex.

late, the visual cortex and the superior colliculus stood out much more markedly than those structures in animals which were kept in darkness during the study.

This kind of work has been further explored by Ingvar (25), who has begun to make precise measurements of the oxygen consumption of the cerebral cortex on the basis of the measurements of local blood flow coupled with measurements of oxygen tension.

Another area which has been of some interest is the difference in oxygen requirements between neuron and glia. Helen Hess (22) recently has made some interesting measurements from which she has concluded that the neuron has a considerably greater oxygen demand than do the glial cells. On the basis of careful counts of neuronal and glial density in the areas of the cortex with which she worked, she has calculated that 90 to 95 percent of the oxygen consumption of the cortex is attributable to the neurons and their processes, with relatively insignificant amounts being utilized by the glial cells even though the glial cells exist in larger numbers. Hydén (24) has recently differentiated metabolically between the neurons and glia in the nucleus of Deiter. He studied certain enzymes of the oxidative cycle, succinic dehydrogenase and cytochrome oxidase, before and after vestibular stimulation. He found that following stimulation there is a marked increase in these enzymes in the neurons contrasted with their decrease in the oligodendroglia which are associated with these neurons. From this he concluded that the glial cells exert a sparing effect upon the neurons, that is, they give priority to the neurons in terms of adaptation to oxygen need.

It has been known for some time now on the basis of the work in the laboratories of Waelsch (49) and of Richter (39) that there is an active protein turnover in the brain. Within the past few years this protein turnover has been studied from the regional point of view. MacLean (35) has made very interesting autoradiograms of the brain following the administration to animals of methionine tagged with radioactive sulfur. These autoradiograms indicate those parts of the brain which have a high turnover rate for methionine, or at least which incorporate methionine at a high rate. They have shown, interestingly enough, that the hippocampus seems to stand out from the rest of the brain for its extremely high ability to incorporate methionine, which presumably indicates a very high rate of protein turnover. Lajtha (31) in Waelsch's laboratory has studied protein turnover rates in various parts of the brain and has demonstrated that these show a differential pattern.

What about the question of relating regional metabolism to function? I have already mentioned the observations of Sokoloff, in which blood flow (presumably an index of oxygen consumption) can be shown to be associated with what is expected to be alteration in local functional activity. I should like to mention also the classical work of Flexner (16), who correlated the appearance of certain enzymes in the brain with functional maturation. He was able to correlate the first appearance of the EEG in embryonic guinea pigs with a rise in enzyme systems which are related to oxidative metabolism, such as ATPase, succinic dehydrogenase and cytochrome oxidase. More recently, Friede (17) has shown some interesting relationships of structural organization in the case of these very same enzymes. He has measured the gradations in succinic dehydrogenase and cytochrome oxidase, which occur in the various nuclei of the thalamus, and has shown that these same gradients occur in the cortical areas to which these nuclei are known to relate anatomically.

Perhaps the most exciting regional neurochemistry of recent years has involved the biogenic amines. In 1943 Raab (38) first described the presence in the brain of pressor substances which he further identified as catecholamines. In 1954 Martha Vogt (48) in Gaddum's laboratory made measurements of norepinephrine distribution in the brain. Patterns of distribution for serotonin have also been developed (5). Although norepinephrine and serotonin have occupied the center of the stage with respect to the biogenic amines, it is only fair to point out that these are by no means the only amines which exist in the brain, nor are they the only amines of possible significance. One very interesting such substance is gamma-aminobutyric acid (GABA) discovered in the brain by Eugene Roberts (40). GABA is a substance which can be called an amine and which exists in the brain and apparently nowhere else. It is differentially distributed within the brain and may have some relationship to function. When Roberts discovered GABA in the brain no one really knew what to do with it. From quite a different approach, an inhibitory factor was also discovered in nervous tissue by Bazemore, Florey and Elliott (4). They proceeded to fractionate their extracts and to purify and crystallize the activity. When they had finished they found that they had a crystalline substance which had a great deal of inhibitory activity and which they identified as gamma-aminobutyric acid.

Histamine has been shown by Adam (1) to have an important place in the brain and to be differentially distributed within that organ in approximately the same regions which are high in other amines, except that histamine appears to be highest in the hypophysis. Axelrod (2) has shown the presence in the brain of an enzyme capable of inactivating histamine, an enzyme which he calls imidazole-N-methyl-transferase. This enzyme is not only high in the brain and differentially distributed there but has its highest concentration in the hypophysis, as has histamine. Dopamine has also been shown to be an amine present in the brain, usually in association with norepinephrine of which it is the precursor. It is possibly related to some of the functions of norepinephrine (15).

Why is there so much interest in these biogenic amines

in recent years? In the first place, these amines are active pharmacological substances with potent properties elsewhere in the body. It is therefore reasonable to assume that if they exist in the brain they must have some function there. Secondly, they are provocatively distributed; that is, their distribution is not uniform throughout the brain. Its concentration in certain areas provokes speculation. These are the areas which the neurophysiologists have suggested may have something to do with effect, mood and emotional state. Perhaps more important than any of the above is the fact that these amines show remarkable changes in concentration in the brain with drugs which are known to affect mental state and behavior. Perhaps the first of these relationships was pointed out by Gaddum (18) when he showed that serotonin, which exists in the brain in this differential pattern, has the remarkable property of being inhibited with respect to its effects on smooth muscle by lysergic acid diethylamide (LSD-25). This substance has been of considerable interest because of its ability to produce a toxic hallucino-genic psychosis. The relationship which Gaddum pointed out highlighted the possible relevance of serotonin to the action of this drug and, in fact, to subjective mental states generally.

Following on the heels of that important discovery was the finding by Shore and his associates (43) in Brodie's laboratory that reserpine, the potent tranquilizing agent, was associated with a fall in the level of serotonin and certain other amines in the brain. A third drug known to affect mental state is iproniazid, which Zeller (50) showed to be a potent monoamine oxidase inhibitor. Iproniazid by that time had been discovered to be a euphoriant and was in fact abandoned for the treatment of tuberculosis because of its ability to cause excessive euphoria, agitation and even psychosis. The finding that this agent was also a potent monoamine oxidase inhibitor awakened a great deal of interest in it, in its possible relationship to the amines and, in turn, their relationship to excitement and mood.

What do we know about the role of specific biogenic amines in normal behavior and in the action of phrenotropic drugs? Unfortunately, the answers are not yet in, and all we have for the present are some rather vague generalizations. Without attempting to review a very active field in a rather inadequate way, I should like to mention a few of the generalizations which I think are possible. I think it is fair to conclude that reserpine sedation is somehow related to the release of amines from the brain. Some of the evidence for that lies in the generally good correlation between sedative properties and central amine release in a large series of congeners of reserpine (7). But which amine is the one which is responsible? Here the evidence is not so clear. There is some suggestion that it may be serotonin. On the other hand, for example, if one sedates an animal with reserpine, thus lowering the brain serotonin level, and then raises that level by the administration of a precursor of serotonin, 5-hydroxytryptophan, one does not reverse the sedation. Such sedation, however, can be reversed by the administration of DOPA, a precursor of norepinephrine (9). But there are many arguments against norepinephrine as the amine specifically involved in reserpine sedation. At the present time, any evidence which implicates one or another amine as being the one responsible for reserpine sedation is not conclusive (30).

What about the excitement produced by the monoamine oxidase inhibitors? Can this be attributed to their inhibition of this important enzyme? I think the evidence for this conclusion is fairly good. By testing a large number of compounds which are monoamine oxidase inhibitors, it was found that generally there is a good relationship between their ability to inhibit this enzyme in vivo and in vitro and their ability to produce behavioral effects similar to excitement or stimulation. But again, when one tries to define the particular amine, the increase in which is associated with or responsible for the behavioral change, one finds a mass of conflicting evidence. There are data compatible with the hypotheses that this amine is serotonin, and data, on the other hand, that suggest it is norepinephrine which is specifically related to the stimulant effects of monoamine oxidase inhibitors. That is to say, there is equally good evidence which is incompatible with one or the other view (6, 8, 22). Recently there has been new evidence to suggest that the specific amine may be dopamine. Carlsson (10) and Everett (14), working independently, have shown a good relationship between cerebral levels of dopamine and behavioral stimulation following monoamine oxidase inhibitors, alone or in association with DOPA, the precursor of dopamine.

What are some of the possible sources of confusion in relating the central amines to behavior? In the first place, although this field is in fact an area of regional neurochemistry, one could say that perhaps it has not been regional enough. Much of the work with amines and the various drugs has involved measurement of the concentration of these agents in the brain as a whole, or, if in parts of the brain, in relatively gross parts like the hypothalamus. Furthermore, we do not know which is the active form of the amines. We know that the amines exist in free form and in bound form, and there are many speculations that one or the other is the active form. If one were to try to relate the whole body content of serotonin, let us say, to mental state, it would be pointed out immediately and quite validly that more than 95 percent of the serotonin in the body is present in the gastrointestinal tract and is, therefore, quite irrelevant to behavior. To what extent are we studying the relevant concentrations of serotonin if we study the amount of serotonin present in the brain or even in the hypothalamus? What we would really like to know is the effective concentration of each amine at its site of action. Until we know what the effective form of the agent is and specifically where it acts, the correlations which we can develop will be open to the criticism that they are irrelevant, fortuitous and perhaps spurious.

There is also the criticism that much of the behavioral characterization of the action of these amines has been made by groups that are extremely sophisticated in biochemistry or in pharmacology but less sophisticated in the behavioral sciences. What is described as excitement or sedation, for example, in some of the literature is difficult to characterize further. One does not know whether the excitement is due to some endogenous effect in the brain or due to a pain in the gut. One does not know whether the sedation which is described is the result of muscular paralysis, general malaise or lack of motivation. There is at the present time an increasing awareness of this problem and the beginnings of collaboration of experts in biochemistry and pharmacology with experts in experimental psychology. From this kind of interaction we can expect a great deal more valid and fruitful activity.

These comments serve to reinforce our recognition of the importance of the overlap of neurochemistry with various disciplines. The relatively new area of regional neurochemistry represents an interface of neurochemistry with the anatomical sciences. But there are other overlaps of neurochemistry, representing other frontiers which are presently of considerable interest. There are the conjunctions of neurochemistry with physiology, with pharmacology, with endocrinology, with immunochemistry. As an example of current work which represents one of these liaisons, I should like very briefly to mention the work of Harris, Scott and Michael (20). They have shown the presence in the hypothalamus of localized chemoreceptors that are sensitive to the presence of circulating estrogen capable of triggering off sexual behavior in an animal normally not susceptible to such behavior.

I would like to mention some very interesting work which is going on in the borderland among neurochemistry, embryology and immunochemistry. That is the work of Rita Levi-Montalcini at Washington University (33). She has, in a very ingenious series of experiments, demonstrated the presence of growth-promoting substances for certain types of nervous tissue in sarcomas, then in snake venom and finally in salivary gland extracts. These growth promoting substances are highly specific for nervous tissue and, in fact, appear to involve the sympathetic nervous system more than anything else. Even more interesting than this is that she has been able to prepare antisera to this growth substance which, when injected into developing fetal animals, suppress the growth of the sympathetic nervous system. She has been able to produce adult animals which have no sympathetic nervous system, or sympathetic systems which are so atrophic as to be functionally and even anatomically absent. Such animals have interesting pharmacological responses, and their continued study from that point of view and from the point of view of physiology offers some remarkable advantages.

I should like to suggest what may be some of the next frontiers of neurochemistry. The fundamental mechanisms involved in the nerve impulse and in synaptic transmission are areas which are just beginning to be investigated and which, I think, will be successfully elucidated only as the result of an interaction among neurochemistry, neuroanatomy, neurophysiology and pharmacology. The chemical embryology of the nervous system is a field of considerable challenge, for example, the problem of chemical specificity. Why is it that the nerve fibers growing out of the central nervous system recognize the end organs for which they are destined and which they will seek out even when they are transplanted to unlikely parts of the body, as has been demonstrated by the elegant work of Weiss and of Sperry (46)? This is probably on the basis of some chemical specificity which remains to be elucidated and to which perhaps the work of Levi-Montalcini may provide a clue.

Then, of course, there is the very fascinating problem of memory, elaborated by Professor Fessard, which may have an important biochemical component. For a long time there has been speculation that the memory trace may be some unstable chemical compound. The knowledge (which is fairly well established in psychology) that memories, unless otherwise reinforced, tend to decay along an exponential curve suggests a material counterpart in the form of an unstable chemical molecule, the dissociation of which follows a similar curve. In recent years there has been considerably more evidence which not only supports the hypothesis that a chemical molecule may be involved but has provided some basis for speculating about what it may be. The nucleic acids and especially RNA, although DNA may be a better possibility, are being more and more implicated as the chemical basis for memory. I should like to cite two recent experiments which suggest that ribose nucleic acid may somehow be involved in the memory process. One is some interesting work which was done by Sporn and Dingman (12) while they were still medical students at the University of Rochester. They knew that an antimetabolite, 8-azaguanine, had been shown to inhibit RNA synthesis by incorporating itself into the RNA molecule and preventing the formation of the normal substance. They demonstrated that this inhibitor, when administered parenterally, gets into the RNA of the brain rather rapidly and achieves its highest concentration about 30 minutes following its administration. They found that when trained rats were given this compound, these animals, during the period of maximum RNA inhibition, were much less capable of learning new tasks than they had been before, even though they could reproduce a previously learned task at the very same time. In other words, the rats could go through a maze which they had previously learned while they were still under the influence of this drug, but they could not learn a new maze. Then there is the interesting work with cats which has been done by Tschirgi and John* at UCLA, and more recently by Corning and John (11) at the University of Rochester upon planaria, the small flat worms which are able to regenerate themselves if cut in half. In this process of regeneration they appear to be capable of transmitting stored information previously taught them to the new half. Such animals, or such halves of animals, do not transfer this information to the regenerated whole animal if they are allowed to regenerate in an environment containing a low concentration of ribonuclease, an enzyme which inhibits the synthesis of ribose nucleic acid.

These are just the vaguest hints that ribose nucleic acid may have something to do with memory. There are, of course, many alternative hypotheses which have to be ruled out and, I suppose, a well defined and thoroughly substantiated chemical theory of memory is still a long time off. But I think that it is quite evident that neurochemistry today is in a very productive stage of its development, aware as it is of its implications to all of the other disciplines of the nervous system. As long as this awareness and interaction can be kept alive, as I am sure it will be kept alive and nourished in such institutes as the Brain Research Institute here, we may predict a very rewarding and productive era.

REFERENCES

1. Adam, H. M., Histamine in the central nervous system and hypophysis of the dog, in Kety, S. S., and J. Elkes, eds., *Regional Neurochemistry*, Proc. 4th Int. Neurochem. Symp., pp. 293–306. London, Pergamon, 1961.
2. Axelrod, J., P. D. Maclean, R. W. Albers, and H. Weissbach,

* John, E. R., B. Wenzel, and R. Tschirgi. Unpublished observations described in Corning and John (11).

Regional distribution of methyl transferase enzymes in the nervous system and glandular tissues, in Kety, S. S., and J. Elkes, eds., *Regional Neurochemistry*, Proc. 4th Int. Neurochem. Symp., pp. 307–11. London, Pergamon, 1961.
3. Bakay, L., Dynamic aspects of blood-brain barrier, in Richter, D., ed., *Metabolism of the Nervous System*, Proc. 2d Int. Neurochem. Symp., pp. 136–520. London, Pergamon, 1957.
4. Bazemore, A. W., K. A. C. Elliott, and E. Florey, Isolation of factor I, J. Neurochem., 1:334–39, 1957.
5. Bogdanski, D. F., H. Weissbach, and S. Udenfriend, The distribution of serotonin, 5-hydroxytryptophan decarboxylase, and monoamine oxidase in brain, J. Neurochem., 1:272–78, 1957.
6. ——, Pharmacological studies with the serotonin precursor, 5-hydroxytryptophan, J. Pharmacol. Exp. Therap., 122:182–94, 1958.
7. Brodie, B. B., K. F. Finger, F. B. Orlans, G. P. Quinn, and F. Sulser, Evidence that tranquilizing action of reserpine is associated with change in brain serotonin and not in brain norepinephrine, J. Pharmacol. Exp. Therap., 129:250–56, 1960.
8. Brodie, B. B., S. Spector, and P. A. Shore, Interaction of monamine oxidase inhibitors with physiological and biochemical mechanisms in brain, Ann. New York Acad. Sci., 80:609–16, 1959.
9. Carlsson, A., M. Lindqvist, and T. Magnusson, 3,4-dihydroxyphenylalanine and 5-hydroxytryptophan as reserpine antagonists, Nature, 180:1200, 1957.
10. ——, and B. Waldeck, On the presence of 3-hydroxytyramine in brain, Science, 127:471, 1958.
11. Corning, W. C., and E. R. John, Effect of ribonuclease on retention of conditioned response in regenerated planarians, Science, 134:1363–65, 1961.
12. Dingman, W., and M. B. Sporn, The incorporation of 8-azaguanine into rat brain RNA and its effect on maze learning by the rat: an inquiry into the biochemical basis of memory, J. Psychiat. Res., 1:1–11, 1961.
13. Elliott, K. A. C., Brain tissue respiration and glycolysis, *The Biology of Mental Health and Disease*, 27th Ann. Conf. Milbank Memorial Fund, pp. 54–73. New York, Hoeber, 1952.
14. Everett, G. M., Cerebral amines and behavioral state: a critique and new data, Biochemical Pharm. (in press).

15. ——, Some electrophysiological and biochemical correlates of motor activity and aggressive behavior, in Rothlin, E., ed., *Neuropsychopharmacology*, II:479. Amsterdam, Elsevier, 1961.
16. Flexner, L. B., Enzymatic and functional patterns of the developing mammalian brain, in Waelsch, H., ed., *Biochemistry of the Developing Nervous System*, Proc. 1st Neurochem. Symp. New York, Academic Press, 1955.
17. Friede, R. L., Thalamocortical relations reflected by local gradations of oxidative enzymes; with some notes on patterns of enzyme distribution in nerve cells, in Kety, S .S., and J. Elkes, eds., *Regional Neurochemistry*, Proc. 4th Int. Neurochem. Symp., pp. 151–59. London, Pergamon, 1961.
18. Gaddum, J. H., Antagonism between lysergic acid diethylamide and 5-OH tryptamine, J. Physiol., 121:15P, 1953.
19. Goodale, W. T., M. Lubin, H. E. Eckenhoff, J. H. Hafkenschiel, and W. G. Banfield, Jr., Coronary sinus catherization for studying coronary blood flow and myocardial metabolism, Am. J. Physiol., 152:340, 1948.
20. Harris, G. W., R. P. Michael, and P. P. Scott, Neurological site of action of stilbesterol in eliciting behavior, in Wolstenholme, G., and C. O'Connor, eds., *Neurological Basis of Behavior*, CIBA Foundation Symposium, pp. 236–51. London, Churchill, 1958.
21. Hess, H. H., The rates of respiration of neurons and neuroglia in human cerebrum, in Kety, S. S. and J. Elkes, eds., *Regional Neurochemistry*, Proc. 4th Int. Neurochem. Symp., pp. 200–13. London, Pergamon, 1961.
22. Hess, S. M., W. Doepfner, and S. Udenfriend, Lack of correlation between excitement and brain amine content after monoamine oxidase inhibitors and amine precursors in rats, Pharmacologist, 1:83, 1959.
23. Himwich, H. W., *Brain Metabolism and Cerebral Disorders*. Baltimore, Williams and Wilkins, 1951.
24. Hydén, H., and P. Lange, Differences in the metabolism of oligodendroglia and nerve cells in the vestibular areas, in Kety, S. S., and J. Elkes, eds., *Regional Neurochemistry*, Proc. 4th Int. Neurochem. Symp., pp. 190–99. London, Pergamon, 1961.
25. Ingvar, D. H., Measurements of regional gaseous metabolism and blood flow in the cerebral cortex, in Kety, S. S., and J. Elkes, eds., *Regional Neurochemistry*, Proc. 4th Neurochem. Symp., pp. 118–25. London, Pergamon, 1961.

27. Kety, S. S., Circulation and metabolism of the human brain in health and disease, Am. J. Med., 8:205-17, 1950.
28. ——, The general metabolism of the brain *in vivo*, in Richter, D., ed., *Metabolism of the Nervous System*, Proc. 2d Int. Neurochem. Symp. (1956), pp. 221-37. London, Pergamon, 1957.
29. ——, Measurement of local blood flow by the exchange of an inert, diffusible substance, in Bruner, D., ed., *Methods in Medical Research VIII*, pp. 223-27. Chicago, Year Book Publishers, 1960.
30. ——, Amino acids, amines, and behavior, *Structure and Metabolism of the Nervous System*, Assoc. Res. in Nerv. Ment. Dis., (in press).
31. Lajtha, A., Exchange rates of amino acids between plasma and brain in different parts of the brain, in Kety, S. S., and J. Elkes, eds., *Regional Neurochemistry*, Proc. 4th Int. Neurochem. Symp., pp. 19-24. London, Pergamon, 1961.
32. Landau, W. M., W. H. Freygang, L. P. Rowland, L. Sokoloff, and S. S. Kety, The local circulation of the living brain; values in the unanesthetized and anesthetized cat, Trans. Am. Neurol. Assoc., 80:125-29, 1955.
33. Levi-Montalcini, R., and P. U. Angeletti, Biological properties of a nerve-growth promoting protein and its antiserum, in Kety, S. S., and J. Elkes, eds., *Regional Neurochemistry*, Proc. 4th Int. Neurochem. Symp., pp. 362-77. London, Pergamon, 1961.
34. Lowry, O. H., N. R. Roberts, M. L. Wu, W. F. Hixon, and E. J. Crawford, The quantitative histochemistry of brain II, enzyme measurements, J. Biol. Chem., 207:19, 1954.
35. MacLean, P. D., S. Flanigan, J. P. Flynn, C. Kim, and J. R. Stevens, Hippocampal function: tentative correlations of conditioning, EEG, drug and radioautographic studies, Yale J. Biol. Med., 28:380-95, 1955.
36. Mosso, A., The temperature of the brain, especially in relation to psychical activity, Proc. Roy. Soc. (London), 51:83, 1892.
37. Pope, A., Brain enzymes in mental disease, in Folch-Pi, J., ed., *Chemical Pathology of the Nervous System*, pp. 388-402. London, Pergamon, 1961.
38. Raab, W., Adrenalin and related substances in blood and tissues. Biochem. J., 37:471-73, 1943.
39. Richter, D., and M. K. Gaitonde, The metabolism of 35 S-methionine in the brain, in Richter, D., ed., *Metabolism of the Nervous System*, Proc. 2d Int. Neurochem. Symp. (1956), pp. 449-55. London, Pergamon, 1957.

40. Roberts, E., Formation and utilization of gamma-aminobutyric acid in brain, in Korey, S. R., and J. Nurnberger, eds., *Neurochemistry*, pp. 11–26. New York, Hoeber, 1956.
41. Schmidt, C. F., S. S. Kety, and H. H. Pennes, The gaseous metabolism of the brain of the monkey, Am. J. Physiol., 143:33–52, 1945.
42. Serota, H. M., and R. W. Gerard, Localized thermal changes in cat's brain, J. Neurophysiol., 1:115–24, 1938.
43. Shore, P. A., A. Pletscher, E. G. Tomich, A. Carlsson, R. Kuntzman, and B. B. Brodie, Role of brain serotonin in reserpine action, Ann. New York Acad. Sci., 66:609–17, 1956.
44. Sokoloff, L., Local cerebral circulation at rest and during altered cerebral activity induced by anesthesia or visual stimulation, in Kety, S. S., and J. Elkes, eds., *Regional Neurochemistry*, Proc. 4th Int. Neurochem. Symp., pp. 107–17. London, Pergamon, 1961.
45. ——, and S. S. Kety, Regulation of cerebral circulation, Physiol. Rev. (Suppl. 4) 40:38–44, 1960.
46. Sperry, R. W., Problems in the biochemical specification of neurons, in Waelsch, H., ed., *Biochemistry of the Developing Nervous System*, pp. 74–84. New York, Academic Press, 1955.
47. Thudichum, J. W. L., *A Treatise on the Chemical Constitution of the Brain*. London, Balliere, Tindall and Cox, 1884.
48. Vogt, M., Distribution of adrenaline and noradrenaline in the central nervous system and its modification by drugs, in Richter, D., ed., *Metabolism of the Nervous System*, Proc. 2d Int. Neurochem. Symp. (1956), pp. 553–65. London, Pergamon, 1957.
49. Waelsch, H., Metabolism of proteins and amino acids, in Richter, D., ed., *Metabolism of the Nervous System*, Proc. 2d Neurochem. Symp., (1956), pp. 431–48. London, Pergamon, 1957.
50. Zeller, E. A., J. Barsky, and E. R. Berman, Amine oxidases. XI. Inhibition of monoamine oxidase by 1-isonicotinyl-2-isopropylhydrazine, J. Biol. Chem., 214:267–74, 1955.

PERCIVAL BAILEY, M.D.

Modern Trends in Neuropathology

Dogmatism has no place in science, and dogmatism about the unknown is especially reprehensible. We live by faith, faith in the order of nature, faith in ourselves, and faith in our fellow men. This faith is our most prevalent motivation, and it is a reliable guide for behavior just in so far as it is founded on knowledge. Where knowledge is lacking we may extrapolate with due regard for the uncertainties arising from the incompleteness of our knowledge. The mystics too often neglect this caution. The naturalists must not, and they find within the bounties of nature ample scope for their best endeavors and for the satisfaction of their highest aspirations.—CHARLES JUDSON HERRICK

ABOUT ten years ago failing vision made me abandon neurosurgery and interfered seriously with my use of the microscope. I returned to an early love and have since been primarily preoccupied with psychopathology (9, 11). This preoccupation will be evident throughout this essay. In the course of the last decade I have been aware that many new developments have occurred in the field of microscopic neuropathology. I have, however, used all the time that I was permitted by my ophthalmologist to spend with the mi-

PERCIVAL BAILEY, M.D., is Director of Research, Department of Mental Health, Illinois State Psychiatric Institute, Chicago, Illinois.

croscope to complete the collaborative work I began long ago with von Bonin on the primate neocortical pattern. This work has just been sent off to the publisher (122), complete except, as Dr. Magoun complained, for the cortex of God. This task, we felt, we had neither the strength nor the competence to tackle. Besides, we were not sure that He could properly be classified as a primate.

However that may be, this was my situation when the invitation came to participate in this symposium. My first impulse was to decline on a plea of ignorance, but the temptation was too great. I selfishly accepted, not only because it would force me to renew my acquaintance with a discipline which I had formerly cultivated but also because I am firmly convinced that neuropathology must be the fundamental basis of much of psychopathology. For although many psychiatric syndromes, such as hysteria, can be understood in terms of a disorder of the psychosocial structure (116), others, such as delirium and dementia, cannot. In the past, whenever neuropathology threw light on some abnormal behavior theretofore explained psychogenetically, the "spiritually minded" psychiatrists always cried, "Ah, yes! but you cannot explain thus so and so, so and so, and so and so," as though ignorance were proof of anything but ignorance. We must not forget that the brain is the organ of mentation and anything which affects the brain adversely must necessarily affect that aspect of its functioning which we call the mind. The prophet of the psychogeneticists, Sigmund Freud, knew this, and it is regrettable that his disciples have developed such a biophobia in these later years.

Very little time was needed to convince me that neuropathology is a thriving discipline even in the restricted sense of pathological anatomy, although the centers of its cultivation have moved out of the mental hospitals and institutes. This evolution has been primarily due to the present trend in psychiatry, which makes it almost impossible for a physician to become a professor of psychiatry without being certified by a psychoanalytic institute, and therefore, if he is not biophobic, at least he is bionegligent. The psychogenic orientation of the teachers is reflected in the staffs and budgets of the mental hospitals. Fortunately the drive to understand the brain and its functioning is not confined to the psychiatrist; new technics, as they are developed, are applied to it everywhere.

In making these statements I am well aware that we must look beyond the structural alterations in the brain to those factors which produced them, even though, as Gerard put it, "no twisted thought without a twisted molecule." I do not forget, as von Bonin has said, "structure should be understood as an enduring order impressed upon a flow of energy." The attempt to study structure apart from function is not apt to be very revealing.

The study of neuropathology is not motivated primarily by the desire to understand the etiology of abnormal mental behavior. Its main motive is the typical primate curiosity concerning any natural phenomenon. When physicians began to dissect the human body, at least since the days of Vesalius and Leonardo da Vinci, they noted not only the normal structure of the nervous system but also the abnormal. They could see with their naked eyes such gross alterations as hydrocephalus, microcephaly, encephalocoeles and tumors. However neuropathology, in the customary meaning of the word, did not get under way until the perfection of the compound microscope and the invention of differential staining of the elements of tissues by Sorby in 1860. Virchow's conception of cellular pathology was applied to the brain by a long line of investigators, such as Weigert, Bielschowsky, Nissl, Alzheimer, Jakob and Spielmeyer. Although the elements of the nervous tissue had been seen before by Ranvier and others, ingenious staining methods made these elements more clearly visible. They were demonstrated in startling silhouette by Golgi, Cajal and their pupils, notably Rio-Hortega, in metallic impregnation. The metallic methods are still used fruitfully by Scharenberg and the Scheibels, and the German methods are still the standby of every neuropathological laboratory. Advantageous modifications continue to be made. The combined method of Klüver and Barrera (70), for example, greatly reduces the routine technical work.

The ability of these classic histological methods to reveal structure was limited by the resolving power of the optical microscope which reached its maximum use about 1890. Ultraviolet light increased its resolution only by a factor of less than two. Electrons enable one to reach a resolving power a hundred times that of light. It is obvious that a method which can make visible particles no more than a few atoms in diameter will reveal a wealth of structural detail not previously known to us. This makes us dream that we may one day be able to see the structural arrangements of atoms and molecules so imaginatively but hypothetically depicted by the chemists and physicists. In fact X-ray diffraction technics have drawn them almost into the realm of reality, as in Watson and Crick's double helix model of the structure of deoxyribonucleic acid.

The method of electron microscopy is being actively applied to problems of the structure of all the organs of the body which could not be solved by methods of lesser resolving power, such as ultraviolet, phase-contrast, interference contrast, polarizing or fluorescence microscopy. Important as has been its contributions to anatomy (131), the electron microscope has not been very useful when applied to neuropathological problems. This is not surprising when it is remembered that the first usable electron microscopes were only built between 1931-34, and a satisfactory method of fixation with osmium tetroxide dates only from 1953. The electron microscope has been applied to the study of experimental demyelinization by Webster and Spiro and to the study of the relationship of viruses to the production of tumors (Dmochowski) but, to my knowledge, it has not yet revealed any virus in a neoplasm of the brain. Recently Luse has applied the method to the origin of colloid cysts of the third ventricle, a problem which has intrigued me since the beginning of my scientific career, but she has not solved the problem. In general the meagre results of its application to neuropathological problems have been reviewed up to 1956 by Roizin and Dmochowski (101). Nevertheless, perusal of recent symposia on neuropathology makes it evident that we are at the beginning of an era of restudy of the classic neuropathological disorders with greatly refined methods whose results can be predicted by no one.

It should be evident that the electron microscope will not solve all the problems of neuropathology, for the simple reason that it can be applied only to dead tissue. This was the great defect also of the early microscopical technics used by the classic neuropathologists. It is also a defect of most of the newer microscopical techniques except for phase contrast and other interference microscopes.

In order to study the dynamic processes going on in cells it was necessary not only to have microscopical techniques which would not kill them during the observation but also to be able to isolate them and still to keep them alive. This necessity led to the development of methods of culturing tissues, first applied to the nervous tissue by Harrison in 1907. Hogue succeeded in 1947 in growing neurons with advancing axonal tips. It is now possible to analyze chemically individual nerve cells in culture or in situ, but very little has been learned in this way other than that the protein is composed mostly of globulins (Hydén).

Pomerat began his studies of the neuroglia by this method in 1951. The activity of the glial cells in various pathological processes is being clarified. Their relation to the nerve cells is believed to be very important but, so far, there is little more than speculation that mental disease may be related to the disturbance of chemical processes going on in glial cells as well as in nerve cells. It is believed that glia has lipolytic, glycolytic and proteolytic functions, and that it transports energy-producing substances from the blood stream to the nerve cells, but the evidence is far from conclusive. Hydén notes that, when nerve cells are increasingly active, there is an increased cytochrome-oxidase activity in the glial cells. He suggests that under such conditions the energy-donor function of the glia takes precedence over its supposed transport function.

The methods of tissue-culture have been used not only to study the growth and structure of elements in the nervous system but also have been used greatly to help in conquering infectious diseases which affect it. They have been extensively utilized to maintain and study viruses which proliferate only in living cells, such as the virus of acute anterior poliomyelitis, for which nonnervous tissues from monkey and man are used in vitro. Various other encephalitides are studied by the same methods. Finally vaccines, such as the Salk vaccine, are prepared by their use.

The recent explosion of knowledge in the field of genetics by the study of chromosomes has been greatly aided by tissue culture. It has been established that the normal chromosome number of human somatic tissue is 46 rather than 48. Beginning with the reports of Lejeune in 1959, there have been made remarkable discoveries in the field of sex chromosomes and the disorders to which their aberrations give rise, such as Klinefelter's syndrome or Turner's syndrome. It has also been demonstrated that the mongolian idiot has an extra small autosomal chromosome 21.

It often happens in any branch of science that the invention of a new technic leads to the abandonment of the older before it has yielded its full harvest. This is not happening in neuropathology. Infection, degeneration, neoplasm, intoxication, and the entire gamut of neuropathological disorders, and many exotic new ones such as Kuru disease (Klatzko), from out-of-the-way places are being studied by the classic histopathological technic.

But it seems evident from the general trend of modern research that the understanding of nervous function and its disorders cannot be complete until we get down to the molecular and atomic levels. Perhaps, as Szent-Györgyi (117) maintains, we must penetrate the field of electrons and protons and their energy-transfer systems. This will lead us far away from the traditional pathological anatomy. We cannot yet see what goes on at these levels but we can gain some ideas by inference.

We have, of course, much useful information about the influence of those social factors from the outside which impinge upon the brain and influence much of its functioning, normal and abnormal. Two of these external factors are the organization of society and the influence of teaching and training on the development of the nervous system; but these hardly fall within the realm of neuropathology. Other outside factors act primarily by influencing chemical reactions going on inside the nervous cells, such as radiation—cosmic or other—toxins, nutritive substances and hormones. We may not go so far as Thudichum (120) and say that many forms of insanity will became clear to us when we know the normal chemistry of the brain in its uttermost details yet, if we wish to understand the *why* rather than the *how* of mental disease, I am convinced that we shall have to pursue the problem to its molecular and atomic roots. And the same reasoning applies much more cogently to neurological disorders which are not accompanied by obvious mental disturbance.

This pursuit of the chemistry of the brain to its "uttermost details" is at present in full cry. One aspect which has special relation to neuropathology, as traditionally understood, is the attempt to relate various chemical substances to structures in the cell visible under the microscope. This may be done by causing chemical reactions to take place in given structures, which will result in the formation of colored compounds, and then to observe them under the microscope—a method known as "microcytochemistry." A considerable literature has accumulated concerning the results of such studies. In this way can be located cholinesterase (Koella), oxidative enzymes (Potanos, Wolf and Cowen), ribonucleic acid (Schiff) and nucleoprotein (Feulgen).

Another way of locating chemical substances in given parts of a cell is microdissection, first described by Schmidt in 1859. Modern micromanipulation was greatly developed by Chambers in 1922, but this method is too laborious for general use. In 1930 my old teacher, Robert Russell Bensley, developed a method of obtaining mitochondria (chondrosomes) by emulsification and centrifugation of tissue. In this way it was possible to obtain large enough quantities of these particles to prove that the site of oxidative phosphorylation, the process by which energy of foodstuffs is made available to metabolism of cells and their functions, is located predominantly if not exclusively in them. In this way have been isolated also microsomes—so important for the synthesis of protein in the cytoplasm—nuclei and, in plants, chloroplasts.

The integrative activity of the various structures thus isolated in the central nervous system—whatever the physiological mechanisms by which it is mediated—depends ultimately upon its endogenous supply of energy. Disturbances at this level are often implicated in cerebral dysfunction or alteration. The brain normally derives its energy for synthesis almost exclusively from the oxidation of glucose (49) although ribonucleic acid may form a labile nitrogenpool available as a stopgap source of energy under stress (83). Despite the possible noncarbohydrate sources, energy in the brain comes almost exclusively from *carbohydrate metabolism*.

The rate of oxidation in the brain is among the highest

in the body. Moreover, as Kety has shown, except in very abnormal conditions, it remains remarkably constant, about 3.5 ml. of oxygen per hundred grams per minute. It is this constancy which is essential for the normal functioning of the brain. Abnormal function may be provoked either by a fall in the rate of oxygen-consumption (unconsciousness, coma, convulsions) or by a rise (convulsions). In the absence of exogenous glucose, the brain can respire and survive for only a short time; the glucose-reserve of the brain is exhausted in 10 to 15 minutes. It is oxidized via the intermediate formation of pyruvate whose oxidation is accomplished by operation of the citric-acid cycle (Krebs). This goes on in a complicated series of steps utilizing a long list of enzymes. The relationships of these glucose breakdown-products in the brain with the amino acids present are even more complicated. The pooling of the intermediates resulting from such interaction is obviously a necessary aspect of the normal chemistry of nerve cells.

Nitrogenous substances from elsewhere in the body probably also play a role in carbohydrate metabolism of the brain, for example, as Geiger shows, uridine and cytidine from the liver. Thudichum wrote in 1884: "I could go further and unfold, e.g., a chemical connection between the function of the liver and that of the brain, opening views into the pathology of the future and illuminating, though only with the disappointing brevity of an electric spark, regions as dark as those of general paralysis and melancholy (120)." It would be very interesting to know what was in his mind.

A significant decrease in the level of glucose in the blood is consistently associated with disturbed cerebral and mental behavior. The degree of impairment of consciousness parallels reduction in the consumption of oxygen and therefore the oxidation of glucose. Many studies in insulin-hypoglycemia and coma have shown good correlation between the level of glucose and mental state (Mann). Intravenous injection of glucose promptly restores mental function in insulin-coma. These facts are used in Sakel's treatment of schizophrenia.

Deprivation of oxygen acts selectively on different parts of the nervous system (49), and the results vary with the age of the subject, as Windle has shown. In the adult the brain stem can survive deprivation much longer than the cortex. Even in the cerebrum the neocortex is much more susceptible than the allocortex. In one case which I observed, an adult woman who was asphyxiated during anesthesia for an operation on her thyroid gland never recovered consciousness. She remained in a state known as "akinetic mutism," her visceral functions normal, her sleep-wakefulness pattern normal, but she recognized nobody and made no movements. She was fed by stomach tube and continued to breathe for 18 months. When the brain was removed the neocortex was degenerated throughout; the striatum and pallidum were badly damaged, but the allocortex and brainstem were intact (Figure 1).

The intimate relationship between mental function and the complicated carbohydrate metabolism in its connections with the metabolism of amino acids is well illustrated by two clinical syndromes, beriberi and Korsakov's syndrome. Beriberi is an oriental disease afflicting rice-eating people. It is due to removal of the husks which contain thiamine (vitamin B_1). Diphosphothiamine is needed for

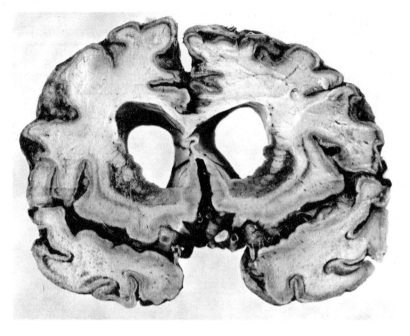

FIGURE 1

Anoxic degeneration of the neocortex cerebri. The allocortex and mesocortex are more resistant.

oxidation of glucose in the brain. Patients suffering from the disease have peripheral palsies and often insomnia, deliria, anxiety, loss of memory and confusion of thought. After treatment with thiamine a Korsakov syndrome (personality change, failure of memory, confabulation, disorientation in space and time) may remain.

Recently in the L. B. Mendel Laboratory of the Elgin State Hospital, Horwitt (18) has produced, by feeding large quantities of unsaturated lipids to chicks depleted of vitamin E, a cerebellitis whose manifestations resemble those of Wernicke's encephalitis, another manifestation of thiamine deficiency. With the present fad for unsaturated

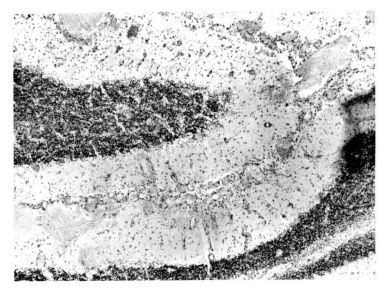

FIGURE 2

Hemorrhagic encephalitis produced in an infant by deficiency of vitamin E. Note the scattered hemorrhages, proliferation of the endothelium of capillaries, and the disappearance of Purkinje cells. (Klüver stain)

vegetable oils, similar hemorrhagic cerebellitis is beginning to be reported in man (Figure 2).

Here is an example of the amazing specificity and individuality of groups of cells in the central nervous system, which must in the final analysis rest on chemical differences as Kety has told you. Deficiency of vitamin B_1 and deficiency of vitamin E are related to similar pathological structural alterations; the former, however, produces them in the paraventricular regions causing a Wernicke's hemorrhagic encephalitis and the latter in the cerebellar cortex with very different symptomatology. Other examples of such specificity might be cited. It is evident that our knowledge of the chemistry of the central nervous system is only in its infancy.

The foundations of "lipid" chemistry of the brain were laid by Thudichum, who isolated and named sphingomyelin, but there is still much to be learned about these compounds. Sphingolipids accumulate in the white matter, almost exclusively in the myelin sheaths. There are a great number of diseases of the nervous system in which demyelination occurs. There is such a disease, swayback, in lambs, which is related to low amounts of copper in the pasture. There is also a long series of heritable disorders of myelination—gargoylism, infantile amaurotic idiocy, Niemann-Pick's disease, Gaucher's disease—which are supposed to result from absence or deficiency of certain enzymes necessary for the conversion of one lipid into another. In pernicious anemia, demyelination of the spinal cord occurs, related to trouble with the absorption of vitamin B_{12}.

Evidence is accumulating that steroids play an important role in metabolism and function of the brain. Therapy with commonly used adrenal steroids may cause mental changes, even psychotic breakdown. The mechanism of their action is unknown, but it is known that in the presence of appropriate enzymes, they can affect systems linked to diphosphopyridine and triphosphopyridine nucleotides.

Proteins and nucleic acids, the most characteristic constituents of living things, may also be affected in pathological processes in the nervous system. The chemistry of these acids has developed in the last 20 years with such rapidity as to bewilder the biochemists themselves. Many of them act as enzymes in combination with simpler structures called "prosthetic groups", obtained from the diet, such as riboflavin or nicotinic acid. Lack of them in the diet may have serious consequences. The oldest known disorder of this sort associated with mental aberrations is pellagra, which was formerly so frequent in mental hospitals. It is known now to be due to deficiency of nicotinic acid in the diet.

One fact which is pregnant with disaster for the nervous system is that the enzymes in the nervous system appear at various times in the development of the organism and vary greatly in concentration, as Flexner has shown (31). Klüver (68) has shown also that there is an ascending porphyrinization of the developing nervous system.

Many serious diseases of infants are now known to result from deviations of enzymatic reactions, such as phenylpyruvic oligophrenia (61). This is a hereditary disorder in which the hepatic enzyme, phenylalanine hydroxylase, is deficient. Another similar disease is galactosemia, which is due to a congenital absence of the enzyme galactose-1-phosphate uridyl transferase necessary for the normal metabolism of galactose. Both conditions result in mental deficiency. Any of the chemical reactions facilitated by enzymes may go astray; more and more of them are being identified.

There was for a time much interest in the metabolism of amino acids. It was found that pyridoxine deficiency in infants resulted in convulsions along with a decrease of gamma-aminobutyric acid, the decarboxylated product of glutamic acid. They are prevented by giving pyridoxamine. Changes in the mental states of epileptic patients who were given glutamine led to the administration with dubious results of glutamic acid to retarded children.

At present, there is great interest in the aromatic amines as substances which may possibly be connected with mental disorder (76). Enzymes concerned with the formation of products derived from tryptamine have marked psychopharmacological effects. Knowledge of the pharmacological inhibitors of this amine metabolism is growing at a rapid rate, promising much understanding which can be utilized for those mental symptoms associated with its disorders. These studies of tryptamine, serotonin and noradrenaline in the brain have focused attention anew on amine oxidase action in the function of the brain. Serotonin and adrenaline are destroyed by monoamine oxidase. Monoamine oxidase in turn is inhibited by iproniazid, a fact which is used in studying the function of serotonin in the brain. The psychomimetic drug, lysergic acid diethylamide, is said to owe its effects on the nervous system to interference with serotonin.

These neuropharmacological studies are even more bewildering than the neurochemical studies, and their intertwinings are comprehensible only to the expert. Although their relationship to the biological reactions which go on normally in the body is often obscure, pharmacological substances may fruitfully be used as tools to elucidate them. Thus prolonged treatment by reserpine produces the most complete form of functional sympathectomy, since the drug has access to sites which cannot be reached by surgical extirpation. Gellhorn (35) has shown how the drug mecholyl can be utilized to elucidate neurophysiological mechanisms.

The presentations of Windle and of Kety make it unnecessary to enter into further details of these chemical aberrations. With all of the complicated chemical reactions going on in the brain, and others elsewhere in the body, it is not surprising that its important function, mentation, is often disturbed. The brain has to be protected from chemicals circulating in the bloodstream; some chemicals gain ready access to the brain while others are more or less excluded. For example, in hepatic precoma, the concentration of ammonia ions in the blood is increased. The result is mental confusion, drowsiness, stupor and finally coma. The attempt to regulate the chemical traffic into and out of the brain is the task of what is called the blood-brain barrier. That there is such a barrier is obvious and experimental attempts to locate it have caused much ink to flow, but there is still no consensus concerning its location.

Experimental technics have been applied to the solution of many problems in neuropathology. Naturally, the effects of deprivation of vitamins come to mind. Disease may be produced not only by deprivation but also by oversupply. Thus recently Auerbach et al. (8) have succeeded in producing phenylketonuric rats experimentally by supplementation of both phenylalanine and tyrosine to the diet. Their work has been confirmed and extended by David Hsia (61) working on a grant from the Mental Health Fund of the State of Illinois.

Another problem which has provoked much experimentation is the causation of the demyelinizing diseases. This is a group of disabling afflictions of the nervous system which have in common these features: destruction of myelin sheaths, conversion of degenerating myelin to cholesterol esters and fats, sparing of axis-cylinders, hyperplasia of astrocytes, distribution in multiple foci, perivenous location, adventitial infiltration of plasma cells, perivascular proliferation of macrophages and absence of pathologic changes in other organs (4).

Attempts to elucidate the causes and mechanism of these changes began more than 20 years ago and are still being actively pursued and are far from completed. The argumentation is based largely on analogy with the results of experiments on animals. The most favored hypothesis is that these disturbances are of an allergic nature due to a hypersensitivity of the nervous tissue produced by immunological mechanisms. It runs somewhat as follows: one or more antigens combine with one or more adjuvants to produce a more potent antigen. Antibodies are produced by plasma cells, circulate in the serum and protect against development of the disease. If a certain threshold of the central nervous system is exceeded, the disease is produced by the encephalitogenic intermediary; and possibly some circulating antigens enter into the reaction which becomes explosive (66).

The analogy with encephalomyelitis following vaccination against rabies in man is very close. In most of the other demyelinating diseases, however, it is necessary to suppose such an autoimmunization as is fairly well-established only for the thyroid gland. After careful comparative study of the pathology of human demyelinating diseases Adams (4) comes to the conclusion that only for three human syndromes is the allergic encephalomyelitis produced in animals sufficiently close to make a similar etiology probable; they are acute necrotizing hemorrhagic leucoencephalitis, acute disseminated encephalomyelitis and acute multiple sclerosis. Schilder's disease and chronic relapsing multiple sclerosis are less probable.

Adams points out that such an etiology in man could be proved only by isolating, from a patient suffering from a demyelinating disease, a substance which would produce the disease in another patient or at least allergic symptoms such as skin sensitivity. Or, a demyelinizing disease could be produced in man by injection of a sterile antigenic substance. The only attempts even remotely related to such proofs, to my knowledge, have been made by Schaltenbrand (106) who, after proper permission had been obtained, injected spinal fluid obtained from monkeys ill with spontaneous encephalitis intracisternally in catatonic and demented patients, producing, in addition to a slight subfebrile reaction, only a transient pleocytosis. Spinal fluid from patients with multiple sclerosis was similarly injected in twelve mental patients, with dubious results. In the absence of such crucial demonstration one is obliged to depend on analogy with the lesions produced in animals other than man. That such studies have raised more problems than they have settled is evident from a recent symposium (66).

Various metals play important roles in the nervous system in combination with substances such as the porphyrins. In general the light metals—sodium, calcium, potassium, iron, magnesium—are important physiologically but the heavy metals—manganese, copper, lead—tend to be toxic. Many of them can now be identified in histological and cytological studies. Many years ago Mella (80) produced degeneration in the basal ganglia of monkeys by administration of manganese. Mercury also causes severe and characteristic changes in the brain, notably atrophy of the striate area of the cortex and of the granular layer of the cerebellum with preservation of the Purkinje cells. The resulting mental syndrome is interesting, marked by timidity, fearfulness, loss of interest in life, later severe depression and loss of memory. From Japan comes a form known as minamata disease, and there has been reported from Iraq severe poisoning by eating bread made from grain treated with a fungicide called Granosan M (Dupont), which contains an organic mercury compound (64).

The effects of intoxication by lead have long been known. It used to be very common in painters but has been eliminated by better preventive measures. It now usually occurs in children from their eating the paint from their toys or in adults from burning old discarded batteries. There is produced in the brain a proliferative endarteritis and thickening of the meninges which cause symptoms difficult to distinguish clinically from intracranial tumor.

Aside from a lamb's disease which is due to the lack of copper in the diet of the ewes, the disease which has undoubtly attracted most attention to the metabolism of copper has been hepato-lenticular degeneration (Wilson's disease). In this disease the copper content of the lenticular nucleus and the liver increases seven to eight times. The disease is a simple recessive hereditary disorder in which there occurs an association of degeneration of the globus pallidus and putamen with cirrhosis of the liver. There is also an increased urinary excretion of amino acids (cystine, glutamic and aspartic acid, arginine and lysine). It is now thought to be due to a deficiency of ceruloplasmin, a copper globulin normally in the blood. Although the increase of copper in the liver is not specific for Wilson's disease, it occurs in no other hepatic disease or neurological disorder.

All of the bewildering array of chemical methods which are now applied to the nervous system are being adapted to make microdeterminations on tissue components (histochemistry) and cell components (cytochemistry) but quantitation of these methods is difficult. Only in a couple of instances semiquantitative methods have been developed, such as the microspectrophotometric measurement of the deoxyribonucleic acid in single nuclei stained by the Feulgen method. A number of these chemical reactions are now well known, such as Schiff's reaction for polysaccharides, Feulgen's reaction for nucleoprotein, and the Koella method for cholinesterase. With the Feulgen method it has been possible to show that deoxyribonucleic acid is exclusively localized in the nucleus and chromosomes (except under special conditions). These methods have been applied to the study of nuclear degeneration (Leuctenberger) and of other pathological processes.

For the most part, the microchemical methods used for the study of the nervous tissue are adaptations of well-known macromethods. The beginning of such adaptation of macromethods was made in 1944 by Lowry and Bessey with their method for riboflavin. They may be colorimetric (ascorbic acid), titrometric (cholinesterase), gasometric (carbon dioxide), dilatometric (peptidase) or microbiological, such as the assay of the rate of growth of a bacterial colony by measurement of its mass. These micromethods enable chemists to deal with substances by other means than test tubes. Who will say that they will not one day study these substances in situ? It is all a matter of developing adequate technics of measurement, which Sinnott has called the "priceless touchstone" of science. One adequate method already in use goes by the intriguing name, the "Cartesian diver."

Although it has been known ever since the discovery of X-rays by Roentgen that radiations can produce pathological changes, it was the devastating effects of the atom bomb which provoked a worldwide study of the biological effects

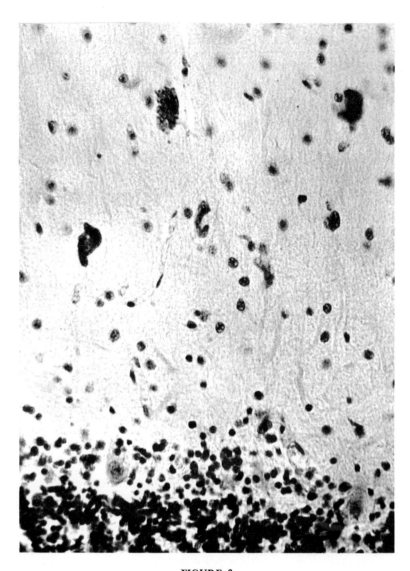

FIGURE 3
Multinucleated glial cells in the cerebellar molecular layer produced by radiation with the betatron. Note the intact capillaries. (Thionin stain)

of ionizing radiation upon the body, including the brain. With atomic explosions being provoked everywhere, it became necessary to know what was the tolerance of the human body to the resulting increase of such radiation in the atmosphere. It is now known that 1000 roentgens is the lethal dose from whole-body exposure for almost all vertebrates. The effect of ionizing radiation is due to absorption of energy from ejected electrons. The ejected electrons cause ionization along their tracks and the biological effects are produced by the ionization. The most disconcerting aspect of this problem is that the evidence of biological change may not be evident for many years.

Beams of electrons can be produced by machines known as "betatrons;" and I have participated in studies of the effects of beta rays produced in this way on the nervous system. My colleagues and I found that the effects of high-energy X-rays on the brain were direct effects and not secondary to vascular occlusion. The cells of the paraventricular and supraoptic nuclei in the hypothalamus of the adult monkey were peculiarly sensitive to this radiation. The normal adult glia cell can be adversely and directly affected (Figures 3 and 4). Others have not agreed that the nerve fibers are more sensitive than the blood vessels, and there is no consensus concerning other details either. It is probable that the effects on the adult brain differ depending upon the source of radiation, its energy, its timing, its nature; whether fast electrons, fast positive ions, electromagnetic radiation of high energy, ultraviolet light or neutrons.

Radiation from isotopes—Co 60, Ta 182, C 14, P 32— has found various uses in experimental procedures. By attaching them to drugs, the distribution of the drugs in

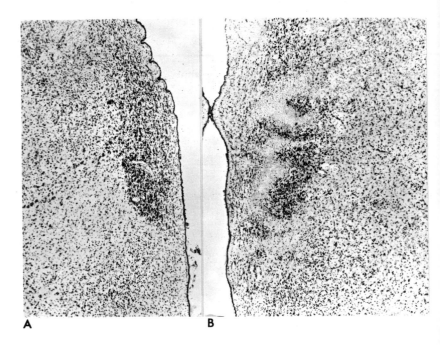

FIGURE 4

Paraventricular nuclei of the hypothalamus of the macaque. A, normal; B, after radiation with the betatron. Note the intact capillaries. (Thionin stain)

the brain can be studied, thus helping to elucidate the mechanisms of their action. They have also been used to create localized destructions in the brain in implants, either with the purpose of studying the functions of deep-lying structures or in the treatment of various disorders of motility. They could also be used in the treatment of mental disturbances by local implantation into nuclei as a substitute for lobotomy.

To me the most appalling result of ionizing radiation is its ability to produce malformations in the developing organism. It is estimated that doubling the present lifetime exposure will double the number of live-born malformed infants. One can convince himself that this prospect is unpleasant from many careful studies, among them the elegant systematic researches of Hicks (45). He has shown how the effects of radiation of the rat with 200 r at various periods of its development produce, depending on the age, different malformations: severe defects of the head at the sixth day of development, defects of the eye at the tenth, defects of the brain and eyes at the eleventh and various abnormalities of the forebrain and cerebellum at the twentieth day. The developmental defects result from differential sensitivity of neural cells, which reach a certain radiosensitive period in their differentiation at varying times in varying locations in the brain. The same differential sensitivity can be shown also to chemical insults. These studies are clearly very important for an understanding of normal cerebral development as well as of malformation, but they are too recent for general principles to be derived from them as yet. Their application to behavioral disturbances is in its infancy.

In the early days of our attempts to relate the normal

development of behavior to the maturation of the brain, Coghill (19) and Herrick (44) spent their scientific lives on this problem, using the *urodele Ambystoma tigrinum*. Although their work convinced the scientific world of the existence of a valid correlation, the animal used was a long way from man in the evolutionary scale. Studies have been made of the development and differentiation of the myelin sheaths in the brain. Attempts to relate the results of these studies to the development of function (121) have not been very instructive because there was no definite correlation between the two variables. Recently Scheibel has followed the neuroanatomical changes in the brains of developing kittens by means of the Golgi method, correlating these findings with the conditioned response to affective stimuli. He found that the time of appearance of this differentiating response correlates with the maturation of axonal plexuses of the cells in the cerebral cortex. With the perfection of chemical technics in our time, a vast amount of information concerning the sequence of chemical differentiation is being acquired (31). Williamina Himwich (50) of the Thudichum Psychiatric Laboratory at the Galesburg State Research Hospital has made many contributions in this field and has recently published a review of the available material. Once these methods have been perfected and focused on specific items of behavior, we will have knowledge which will also illuminate abnormal behavior.

Hooker (56) has studied the development of reflex action in the prenatal period. Studies of the effects of radiation, as we have just noted, demonstrate variable sensitivity at different levels of development and must be correlated with the chemical studies of Flexner. The ontogenesis of enzymes (62) promises to be enlightening; it has been shown by Gordon that the activity of certain enzymes does not manifest itself until the necessary substrate has developed. The various chemical differentiations must, therefore, follow in a definite order, and any alteration in the chain will cause maldevelopment. These chains extend far into postnatal life. It is evident that the effects of physical or chemical injury to the brain duplicate the results obtained by Paul Weiss (127) and other experimental zoologists with surgical extirpations.

The importance of these biological studies of development by chemical, cytological, surgical and radiological methods for the maturation of the human infant has been conclusively demonstrated by Pasamanick (71), Bender (12) and others. Their correlation with the psychological and neurological development as related by Hooker (56), Gesell (36), Piaget (91) and others promises to be much more illuminating than the fanciful oral-anal-genital theme of the dynamic psychologists.

I wish now to say a few words about the application of this new knowledge to that protean group of mental aberrations called "schizophrenia." It is unnecessary to go into biological detail because the excellent summary of Kety (65) is readily available. The group has in common that all of the patients give the impression that they do not have well-integrated personalities, hence the name, schizophrenia. Chaslin called the syndrome *la folie discordante*. Their universal imperviousness to psychotherapy has led even the most eminent psychotherapists—Janet, Freud, Jung—to conclude that the essential defect must be due to a biochemical mechanism. This opinion is supported by much positive evidence since the days when Moreau de

Tours showed how intoxication with hashish aped the syndrome. Since then numerous toxic substances, such as mescal (67) and lysergic acid diethylamide, are known to produce even more similar syndromes.

It is, of course, not supposed that the schizophrenic syndrome is provoked by toxins introduced from without the organism, else they would long ago have been discovered; nor is it surprising that the symptoms provoked by such toxins do not reproduce exactly the spontaneously developing schizophrenic syndrome. At present the most plausible hypothesis is that clearly enunciated by Osmond and Smythies (84), in which it is supposed that toxic substances are formed in the organism by deviation of biochemical reactions. Hoffer (52) seeks them in the metabolism of epinephrine, Heath (42) in the metabolism of globulins in the serum. These deviations are supposed to be genetically determined (Kallmann, 63). That other factors, such as social status, play a precipitating role is not surprising. There are many analogies in medicine, for example, in tuberculosis. Nor is the hypothesis incompatible with the findings of the psychologists. Whether one supposes incongruity of affect and deviation of perception or, as the psychoanalysts would have it, uncertainty about sexual identity and archaic procreative fantasies to be characteristic of the syndrome, it still must be explained why disabling disease develops in only a certain proportion of human beings. Analogy is not proof, but the analogies of the symptoms provoked by toxins are so impressive as to place the burden of proof on those who maintain that such factors play no role in their appearance in schizophrenic patients.

Many are impressed by the failure of histopathologists to demonstrate constant characteristic structural alterations in the brains of deceased schizophrenic patients. We must remember that the search was conducted by the old crude methods. It must be resumed with all the refined, infinitely more revealing, methods which we mentioned in the earlier parts of this essay. There are many indications that the search should be concentrated on the brainstem (108).

In the past 50 years there have been many reports of supposed abnormalities in the metabolites of the blood and urine of schizophrenic patients, but these were never proved to be consistent or characteristic of the disease. One of the latest of such reports was made by Akerfeldt, who suggested that ceruloplasmin in the serum of mental patients might be the cause of its increased oxidizing activity. This was promptly disproved by Horwitt and his co-workers in the L. B. Mendel Laboratory of the Elgin State Hospital. They showed (58) that the positive Akerfeldt test was often a function of the low ascorbic acid level of the diet of institutionalized patients. But the search continues. Recently Sankar has reported that children with a diagnosis of schizophrenia have higher levels of inorganic phosphate in plasma and erythrocytes than control hospitalized children, thus supporting Hoagland's earlier finding in adults. Hoagland reports also that non-schizophrenics have, in their plasma, an inhibiting factor, a small molecule attached to a larger globulin molecule; the absence of which accounts for the disturbing effects of schizophrenic plasma on animal and human behavior.

It seems evident now that the failure, over years of effort, to find consistent metabolic errors by examination of the blood and urine of schizophrenic patients was due in part to the unquestioned heterogeneity of the group which must be broken up into smaller units, more clearly delineated. This has often been attempted; thus recently Meduna (79) has tried to separate out a group which he names "oneirophrenia." The fruitfulness of such partial studies has been demonstrated by Gjessing's (38) isolation and characterization of intermittent catatonia in which he has proved, in an admirable series of investigations, that there is a disturbance of the metabolism of nitrogen. His work has been confirmed and broadened by its application to such cycles in rats by Richter (97).

So much for schizophrenia (and one might make similar remarks about the other psychoses) but what of the *psychoneurotics*—the thousands of weaklings and fools (Freud called them both) who profit most from psychotherapy? They are weak, as Pierre Janet and Weir Mitchell pointed out long ago, and they are foolish in the endless ways which they adopt to overcome their weakness, as Freud and his disciples have analyzed ad infinitum. They are weak because of bad heredity, enfeebling illness, inadequate diet, or merely relatively weak compared to the social problems they must solve. They are also hypersensitive, as Dejerine said. William James called them "tender-minded." Their symptoms may be transitory, following illness, and disappear with no more than the care of a good physician; they may disappear with the social impediments which precipitated them; or they may be permanently recalcitrant to medicine, social betterment and psychotherapy, whether it be in the form of religion, Christian science, deconditioning, counseling or psychoanalysis. It is interesting that Janet in his lectures at the Collège de France in 1924 made the following statement (57): "One little-noticed substitute for religion is destined to do perhaps more than all others to put religion out of style: scientific psychotherapy, which studies and remedies scientifically those states of mental depression for which religion is the sovereign but imperfect popular remedy." It is probable that the psychoneuroses cannot always be fruitfully related to the sort of neuropathology which we have been discussing, and they have undoubtedly far broader implications than the schizophrenias. Their psychological, social and anthropological determinants are talked about enough by others. I wish at this point only to remind you that a careful, biochemical study of many psychoneurotic patients will reveal abnormalities which will save them much time and expense wasted in psychotherapy. Such patients frequently come under my observation.

Now it is evident that the behavior of the human organism, normal and abnormal, cannot be explained exclusively on the basis of chemical reactions. We must take into account also cybernetics (7, 132), learning theory (30) and game-playing (116). I am well aware of the sociological factors, and those internalized social factors which we call psychological (78), which play important roles in the causation of abnormal behavior, mental and otherwise, but I am not presently interested in them. I am one of those "others" of whom Freud spoke when he said that he did not talk of the biological factors because others spoke of them enough.

All explanatory hypotheses based on the social environment—upbringing, teaching, training, interpersonal relations, communication—suppose an intact and normally functioning brain. This supposition is purely gratuitous and must be proved as the first step in the understanding of the behavior of any human being. Such hypotheses have never been adequate by themselves to explain the behavior of patients suffering from any mental disorder, whether depression, compulsion, phobia or schizophrenia. They are effective only on a prepared terrain. This Janet knew, as did Freud and Jung. They expressed their conviction on many occasions. Surely no one will dispute the opinion that the material I have discussed makes it evident in which direction we are to seek for the explanations of mental deficiency. We cannot expect to encounter wolf-children and others kept in asocial captivity very often.

When faced by the bewildering array of facts which the chemists, physicists and biologists have accumulated, one's first reaction is wonder, the same wonder felt by primitive man when he contemplated the starry universe. Today it is the universe of the infinitesimally small which is opened before us, so complicated that only experts can master its details. Out of these facts, order must be brought, as Pavlov (86) warned his students in his last message. This order is not yet apparent and its lack must play a role in the reluctance of many psychiatrists to take these discoveries seriously. Moreover, they hesitate to attempt to gain first-hand information; the physiological laboratory is messy and bloody, the chemical laboratory filled with bad smells, as Robert Hutchins is said to have remarked. How much more pleasant to sit in an air-cooled office and listen to the yarns spun by weaklings and fools. As for the really sick, let them go to the mental hospitals and be cared for by foreigners who are obliged to accept such disagreeable tasks as proper recompense for being admitted to the land of the free and the home of the brave.

Another reason for this reluctance to face the implications of these neuropathological findings is the lack of a comprehensive theory which will relate this material to the preoccupations of the "tender-minded" psychoneurotic, notably, as Jones wrote, "the seat of the soul, the purpose of life, and the control of our animal nature." And the psychotherapists are usually equally tender-minded. They are rarely of the cloth from which scientists are cut. They cannot maintain an attitude of suspended judgment, the earmark of a scientist; they must have a certainty to control their own anxiety. This they can acquire much more easily from the training and theory of psychoanalysis. Generations of psychoanalysts have worked out a comprehensive theory which gives a sense of completeness (if disturbing facts are ignored) just as before them generations of Christian theologians did much better. "A universe explained even by bad reasons is a comfortable universe" (Camus).

"We suffer if we cannot unify our beliefs into a system, particularly if this system is what we live by. The existence of persons who dispute our convictions breaks up this unity and causes great pain, especially to those who are already troubled by doubts. It is the doubter to whom his neighbor's unbelief causes the most acute distress." (57) Call such a system theology, metaphysics, philosophy, dynamic psychology—what you will—a science it is not and never will be.

Care of the fools and weaklings who bedeck the couches of the psychoanalysts is a socially useful task, but one which is not properly medical and can be adequately accomplished by others than by physicians, as Freud knew full well. Psychiatrists, who are by definition physicians, should get on with the task which Freud said is the important problem for the future: the relation of physical to mental states. This problem we can now attack with ever increasing confidence if we remain free from the stifling inhibitions of preconceived systems of thought. We cannot be Einsteins, but we can remember that he said he discovered relativity because he never believed in an axiom. We should take into account only facts which can be measured by scientific method and build our hypotheses on them. We shall not then build artificial edifices of guesswork concerning the structure of the mind, forgetting that functions are not things and cannot properly be said to have structure.

We are far from Thudichum's goal of knowing the chemistry of the brain in all its ultimate details but we are rapidly accumulating intimate knowledge of the chemistry of that organ, and some of this knowledge will give us clues to the etiology of disease of the nervous system. We will then, as Thudichum predicted, be in a better position to develop a scientific therapy of those neurological disorders whose lack of response to the therapies available in my youth drove me to take up neurosurgery and drove Sigmund Freud into the field of neuroses. Today the psychoses are equally recalcitrant to our presently available therapies, including the surgical and psychoanalytical. This would not surprise Freud who, to his last breath, expected the chemists to find the remedies for these stubborn maladies. He told Schilder, on the point of emigration to America, to hurry and study the psychology of schizophrenia before someone discovered an injection and the schizophrenic became as rare as the American Indian. Schilder did not live to use the injection, and perhaps neither shall we, but we have more reason than he to look forward confidently to its realization.

In looking back over the developments which I have mentioned all too briefly in this essay, I am astonished at the number of new technics which were invented in the 1930's and 40's. On them our new knowledge is based. They may bring to an end the final flowering of the dark age of psychiatry when it tried by the power of fantasy, working on hearsay evidence, guided by *ad hoc* hypotheses, and insights hoary with age but distorted, exaggerated and couched in pseudoscientific vocabulary, to solve not only its own legitimate problems but also those of society as well. These insights its adepts now repeat *ad nauseam* like a phonograph stylus unable to get out of a groove.

But the renaissance of psychiatry does not seem imminent. It is at present almost impossible to become professor of psychiatry without having gone through the mystical process of a personal analysis. This places the teaching of psychiatry in the hands of extraacademic organizations, a very unhealthy situation. There is some awareness of this alien influence in the university faculties, evidenced by the attempts to place chemists at the head of psychiatric faculties. This is surely foolish since, however important a role chemical disorder may play in the production of psychoses, chemistry is not psychiatry. The head of a psychiatric service must be a trained clinician. As further evidence of confused thinking one may cite the attempts to have these chemists analyzed. This is even more foolish since psychoanalysis, being nearer to a religion than a science, is no more an adequate substitute for broad psychiatric experience than is chemical competence. Some way must be found to break the grip of the analysts on psychiatric education if the field is to develop healthily.

I have not attempted a review of neuropathology which, being an obvious impossibility within the time at my disposal, I am sure I was not supposed to undertake, for it enters into every nook and cranny of our knowledge of the brain. Nor have I indulged in any presumptuous predictions. I have merely expressed my belief that these recent developments portend results of future investigation on which we can build much better founded theories of normal and abnormal functioning of the brain, provided we do not meanwhile, in our inability to make proper use of our discoveries, try to outdo God, who preserved at least Noah and his children, and either crowd ourselves off the entire earth in a population-explosion or blow ourselves out of His universe. Then such trinitarian pseudosolutions as Superego, Ego, Id will cease to preoccupy us.

REFERENCES *

1. Abood, L. G., Some chemical concepts of mental health and disease, Proc. Assoc. Res. Nerv. Ment. Dis., 37:384–96, 1959.
2. ——, Neuronal metabolism, in Field, J., H. W. Magoun, and V. E. Hall, eds., *Handbook of Psysiology Neurophysiology III*, pp. 1815–26. Washington, D. C., American Physiological Society, 1960.
3. Abramson, H. A., ed., *Transactions, Conference on Neuropharmacology*, Vols. 1–5. New York, Josiah Macy, Jr. Foundation, 1954–59.
4. Adams, R. D., A comparison of the morphology of the human demyelinative diseases and experimental "allergic" encephalomyelitis, in Kies, M. W., and E. C. Alvord, Jr., eds., *"Allergic" Encephalomyelitis*, pp. 183–209. Springfield, Thomas, 1959.
5. Ariëns-Kappers, J., ed., *Proceedings First International Meeting of Neurobiologists*, Amsterdam, Elsevier, 1956.
6. Arnold, A., P. Bailey, and J. S. Laughlin, Effects of betatron radiations on the brain of primates, Neurology, 4:165–78, 1954.
7. Ashby, W. R., *Design for a Brain*, 2d ed. New York, Chapman and Hall, 1960.

* In addition to these sources, I have profited also from numerous conferences with Leo Abood, Eric Brunngraber, Harold Himwich and Max Horwitt. They are not responsible for any errors of fact or emphasis.

8. Auerbach, V. H., H. A. Waisman, and L. B. Wyckoff, Jr., Phenylketonuria in the rat associated with decreased temporal discrimination learning, Nature, 182:871–72, 1958.
9. Bailey, P., The great psychiatric revolution, Am. J. Psychiatr., 113:384–406, 1956.
10. ——, Advances in basic science which have arisen in the psychiatric clinic, in Beecher, H. K., ed., *Disease and the Advancement of Basic Science*, pp. 315–40. Cambridge, Harvard Univ. Press, 1960.
11. ——, A rigged radio interview, Perspectives Biol. Med., 4:199–265, 1961.
12. Bender, L., The brain and child behavior, Arch. Gen. Psychiatr., 4:531–47, 1961.
13. Brady, R. O., and D. B. Tower, eds., *The Neurochemistry of Nucleotides and Amino-Acids; a Symposium*. New York, Wiley, 1960.
14. Brosin, H. W., ed., *Lectures in Experimental Psychiatry*, Pittsburgh, Univ. of Pittsburgh Press, 1961.
15. Brücke, R., ed., *Biochemistry of the Central Nervous System*, Proc. 4th Int. Cong. Biochem. Symp., Vienna (1958), Vol. 3. London, Pergamon, 1959.
16. Cain, A. J., The histochemistry of lipoids in animals, Biol. Reviews, Cambridge Phil. Soc., 25:73–112, 1950.
17. Calvert, D. N., and T. M. Brody, Role of the sympathetic nervous system in CCl_4 hepatoxicity, Am. J. Physiol., 198:669–76, 1960.
18. Century, B., M. K. Horwitt, and P. Bailey, Lipid factors in the production of encephalomalacia in the chick, Arch. Gen. Psychiatr., 1:420–24, 1959.
19. Coghill, G. E., *Anatomy and the Problem of Behaviour*. London, Cambridge Univ. Press, 1929.
20. Cumings, J. N., *Heavy Metals and the Brain*. Oxford, Blackwell, 1959.
21. ——, and M. Kremer, eds., *Biochemical Aspects of Neurological Disorders*. Oxford, Blackwell, 1959.
22. Davidson, C. E., *Consciousness and the Chemical Environment of the Brain*. Columbus, Report of the 25th Ross Pediatric Research Conference, Ross Laboratories, 1957.
23. Dawson, I. M. P., The histology and histochemistry of gargoylism, J. Path. Bact., 67:587–604, 1954.

24. Eccles, J. C., *Neurophysiological Basis of Mind.* Oxford, Clarendon, 1953.
25. Edsall, J. T., The physician and the scientific revolution of our time. Pharos of AOA, 24:162–69, 1961.
26. Elliott, K. A. C., Brain tissue respiration and glycolysis, in *Biology of Mental Health and Disease,* 27th Annual Conf. Milbank Memorial Fund, pp. 54–70. New York, Hoeber, 1952.
27. Elliott, K. A. C., I. H. Page, and J. H. Quastel, eds., *Neurochemistry.* Springfield, Thomas, 1955.
28. Eränkö, O., *Quantitative Methods in Histology and Microscopic Histochemistry.* Boston, Little, Brown, 1955.
29. Errera, M., and A. Forssberg, eds., *Mechanisms in Radiobiology.* New York, Academic Press, 1961.
30. Eysenck, H. J., *The Dynamics of Anxiety and Hysteria; an Experimental Application of Modern Learning Theory to Psychiatry.* New York, Praeger, 1957.
31. Flexner, L. B., Enzymatic and functional patterns of the developing mammalian brain, in Waelsch, A., ed., *Biochemistry of the Developing Nervous System,* Proc. 1st Int. Neurochem. Symp., pp. 281–95. New York, Academic Press, 1955.
32. Folch-Pi, J., and M. Lees, Studies on the brain ganglioside strandin in normal brain and in Tay-Sachs disease, J. Dis. Children, 97:730–38, 1959.
33. Gaffron, H., The origin of life, Graduate Journal, Univ. of Texas, 4:82–133, 1961.
34. Garattini, S., and V. Ghetti, eds., *Psychotropic Drugs.* Amsterdam, Elsevier, 1954.
35. Gellhorn, E., *Autonomic Imbalance and the Hypothalamus.* Minneapolis, Univ. of Minnesota Press, 1957.
36. Gesell, A. L., *The Embryology of Behavior; the Beginnings of the Human Mind.* New York, Harper, 1945.
37. Gibbs, F., ed., *Molecules and Mental Health.* Philadelphia, Lippincott, 1959.
38. Gjessing, R., Beiträge zur Kenntnis der Pathophysiologie des katatonen Stupors, Arch. Psychiatr. Nervenkrank., 96:319–473, 1932.
39. Glick, D., *Technics of Histo- and Cytochemistry.* New York, Interscience, 1949.
40. Gömöri, G., *Microscopic Histochemistry; Principles and Practice.* Chicago, Univ. of Chicago Press, 1952.

41. Heath, R. G., *Studies in Schizophrenia*. Cambridge, Harvard Univ. Press, 1954.
42. —, Physiological and biochemical studies in schizophrenia with particular emphasis on mind-brain relationships, in Pfeiffer, C. C., and J. P. Smythies, eds., *International Review of Neurobiology*, 1:299–331. New York, Academic Press, 1959.
43. —, and B. E. Leach, Multidisciplinary research in psychiatry, in Rado, S., and G. E. Daniels, eds., *Changing Concepts of Psychoanalytic Medicine*, pp. 201-24. New York, Grune and Stratton, 1956.
44. Herrick, C. J., *The Evolution of Human Nature*. Austin, Univ. of Texas Press, 1956.
45. Hicks, S. P., Radiation as an experimental tool in mammalian developmental neurology, Physiol. Rev., 38:337–56, 1959.
46. —, Developmental brain metabolism, Arch. Path., 55:302–27, 1953.
47. —, C. J. D'Amato, and M. J. Lowe, Development of the mammalian nervous system. I. Malformations of the brain, especially the cerebral cortex, induced in rats by radiation. II. Some mechanisms of the malformation of the cortex, J. Comp. Neurol., 113:435–69, 1959.
48. Hierons, R., Changes in the nervous system in acute porphyria, Brain, 86:176–92, 1957.
49. Himwich, H. E., *Brain metabolism and cerebral disorders*. Baltimore, Williams and Wilkins, 1951.
50. Himwich, W. A., Biochemical and neurophysiological development of the brain in the neonatal period, (in press).
51. —, and E. Costa, Behavioral changes associated with changes in concentration of brain serotonin, Fed. Proc., 19 (Pt. 2): 838–45, 1960.
52. Hoffer, A., Adrenochrome and adrenolutin and their relationship to mental disease, in Garattini, S., and V. Ghetti, eds., *Psychotropic Drugs*, pp. 10–20. Amsterdam, Elsevier, 1957.
53. —, Adrenolutin as a psychomimetic agent, in Hoaglund, H., ed., *Hormones, Brain Function and Behavior*, pp. 181–94. New York, Academic Press, 1957.
54. —, and H. Osmond, *The Chemical Basis of Clinical Psychiatry*. Springfield, Thomas, 1960.
55. —, and J. Smythies, Schizophrenia: a new approach; results of a year's research, J. Ment. Science, 100:29–45, 1954.

56. Hooker, D., *The Prenatal Origin of Behavior*. Lawrence, Univ. of Kansas Press, 1952.
57. Horton, W. M., The origin and psychological function of religion according to Pierre Janet, Am. J. Psychol., 35:16–52, 1924.
58. Horwitt, M. K., Fact and artifact in the biology of schizophrenia, Science, 124:429–30, 1956.
59. ——, and P. Bailey, Cerebellar pathology in an infant resembling chick nutritional encephalomalacia, Arch. Neurol., 1:312–14, 1959.
60. Hsia, D. Y.-Y., *Inborn Errors of Metabolism*. Chicago, Year Book Publ., 1959.
61. ——, Genetic errors of metabolism and environmental interaction; a synthesis, Ann. New York Acad. Sci., 91:674–83, 1961.
62. ——, Ontogenetic development of enzyme-systems, in Pruzansky, S., ed., *Congenital Anomalies of the Face and Associated Structures*, pp.134–48. Springfield, Thomas, 1961.
63. Kallmann, F. J., Heredity and eugenics, Am. J. Psychiatr., 116:577–81, 1960.
64. Kantarjian, A. D., A syndrome clinically resembling amyotrophic lateral sclerosis following mercurialism, Neurology, 11:639–44, 1961.
65. Kety, S. S., Biochemical theories of schizophrenia, Science, 129:1528–32, 1590–96, 1959.
66. Kies, M. W., and E. C. Alvord, Jr., *"Allergic" Encephalomyelitis*. Springfield, Thomas, 1959.
67. Klüver, H., *Mescal: the Divine Plant and its Psychological Effects*. London, Paul, Trench, Trubner, 1938.
68. ——, Porphyrins in relation to the development of the nervous system, in Waelsch, H., ed., *Biochemistry of the Developing Nervous System*, Proc. 1st Int. Neurochem. Symp., pp. 137–44. New York, Academic Press, 1955.
69. ——, Porphyrins, the nervous system and behavior, J. Psychol., 17:209–28, 1944.
70. Klüver, H., and E. Barrera, A method for the combined staining of cells and fibers in the nervous system, J. Neuropath. Exp. Neurol., 12:400–03, 1953.
71. Knobloch, H., and B. Pasamanick, Syndrome of minimal cerebral damage in infancy, J. Amer. Med. Ass., 170:1384–87, 1959.

72. Korey, S. R., and J. I. Nürnberger, eds., *Neurochemistry*. New York, Hoeber, 1956.
73. ——, and H. Winograd, Biochemical alterations in a case of Heller's disease, J. Dis. Children, 97:668–75, 1959.
74. Kruger, O., ed., Symposium on catecholamines, Natl. Inst. Health, Bethesda (1958), Pharmacol. Review, 11:233–566, 1959.
75. McIlwain, H., *Biochemistry and the Central Nervous System*, 2d ed. London, Churchill, 1959.
76. Marrazzi, A. S., Synaptic and behavioral correlates of psychotherapeutic and related drug activities, in *Some Biological Aspects of Schizophrenic Behavior*, Ann. New York Acad. Sci., 96:211–26, 1962.
77. Mautner, H., *Mental Retardation; its Care, Treatment and Physiological Bases*. London, Pergamon, 1959.
78. Mead, G. H., *Mind, Self and Society*. Chicago, Univ. of Chicago Press, 1934.
79. Meduna, L. J., *Oneirophrenia, the Confusional State*. Urbana, Univ. of Illinois Press, 1950.
80. Mella, H., The experimental production of basal-ganglia symptomatology in Macacus rhesus, Arch. Neurol. Psychiatr., 11:405–17, 1924.
81. Mellors, R. C., ed., *Analytical Cytology*. New York, McGraw-Hill, 1955.
82. Nachmansohn, D., *Chemical and Molecular Basis of Nerve Activity*. New York, Academic Press, 1959.
83. Nürnberger, J. I., and M. W. Gordon, Effects of brief stress on ribonucleic acids and the labile nitrogen pool of brain and liver in the rat, Int. Rev. Cytol., 2:231–51, 1953.
84. Osmond, H., and J. Smythies, Schizophrenia; new approach, J. Ment. Science, 98:309–15, 1952.
85. Pauling, L. C., and H. A. Itano, eds., *Molecular Structure and Biological Specificity; a Symposium*. Washington, D. C., Am. Inst. Biol. Sciences, 1957.
86. Pavlov, I. P., Bequest of Pavlov to the academic youth of his country, Science, 83:369, 1936.
87. ——, A letter to the youth, in *Selected Works*, pp. 54–55. Moscow, Foreign Languages Publishing House, 1955.
88. Pearse, A. G. E., *Histochemistry; Theoretical and Applied*, 2d ed. London, Churchill, 1960.

89. Pennes, H. H., ed., *Psychopharmacology; Pharmacologic Effects on Behavior.* New York, Hoeber, 1958.
90. Pfeiffer, C. C., Parasympathetic neurohumors; possible precursors and effect on behavior, in Pfeiffer, C. C., and J. R. Smythies, eds., *International Review of Neurobiology,* 1:195–224. New York, Academic Press, 1959.
91. Piaget, J., *The Construction of Reality in the Child.* New York, Basic Books, 1954.
92. Pomerat, C. M., Dynamic neuropathology, in Symposium on the Role of Histochemistry and Histometabolic Role in Study of the Central Nervous System, J. Neuropath. Exp. Neurol., 14:28–38, 1955.
93. Proceedings of the Second International Congress of Neuropathology, Excerpta Med., (Sec. VIII) 8:767–896, 1955.
94. Quastel, J. H., Biochemical aspects of anesthesia, in *Proc. 3d Int. Cong. of Biochem.,* pp. 496–504. New York, Academic Press, 1956.
95. ——, and D. M. J. Quastel, *Chemistry of Brain Metabolism in Health and Disease.* Springfield, Thomas, 1961.
96. Rapports et Discussions 3^{ie} Congress International Neuropathologie, Bruxelles (1957), in Van Bogaert, L., and J. Radermeche, eds., *First International Congress of Neurological Sciences,* Vol. 4. London, Pergamon, 1959.
97. Richter, C. P., Hormones and rhythms in man and animals, Rec. Prog. Horm. Res., 13:105–59, 1957.
98. Richter, D., ed., *Metabolism of the Nervous System,* Proc. 2d Int. Neurochem. Symp. (1956). London, Pergamon, 1957.
99. ——, ed., *Schizophrenia: Somatic Aspects.* New York, Pergamon, 1957.
100. Roberts, E., ed., *Inhibition in the Nervous System and Gamma-aminobutyric Acid; Proceedings of an International Symposium.* New York, Pergamon, 1960.
101. Roizin, L., and Dmochowski, L., Comparative histologic and electron investigations of the central nervous system, J. Neuropath. Exp. Neurol., 15:12–32, 1956.
102. Rosenberg, A., and E. Chargaff, Some observations on mucolipids of normal and Tay-Sachs disease: brain tissue, J. Dis. Children, 97:739–44, 1959.
103. Roth, L. J., and C. F. Barlow, Drugs in the brain, Science, 134:22–31, 1961.

104. Russell, L. B., Genetics of mammalian sex chromosomes, Science, 133:1795–1803, 1961.
105. Salzman, N. P., Animal cell cultures, Science, 133:1559–65, 1961.
106. Schaltenbrand, G., *Die multiple Sklerose des Menschen.* Leipzig, Thieme, 1943.
107. Schmidt, G., et al., The partition of tissue phospholipids by phosphorus analysis, J. Dis. Children, 97:691–708, 1959.
108. Sherwood, S. L., Consciousness, adaptive behaviour and schizophrenia, in Richter, D., ed., *Schizophrenia, Somatic Aspects,* pp. 131–46. London, Pergamon, 1957.
109. Sinnett, E. W., Science and the education of free men, Am. Scientist, 32:205–15, 1944.
110. Spielmeyer, W., *Die Anatomie der Psychosen.* Berlin, Springer, 1930.
111. Symposium on amaurotic family idiocy. Tay-Sachs disease, J. Dis. Children, 97:655–760, 1959.
112. Symposium on brain and behavior, Am. J. Orthopsychiatr., 30: 1–49; 292–330, 1960.
113. *Symposium on Cytology in Human Genetics.* Washington, D. C., Am. Inst. Biol. Sciences, 1959.
114. Symposium on genetics of disordered behavior, East. Psychiatr. Res. Assoc., New York, Diseases of the Nervous System, Monograph, (Suppl.) XXI (No. 2): 9–64, 1960.
115. *Symposium on mental retardation,* Publ. Assoc. Res. Nerv. Ment. Dis., (in press), 1959.
116. Szasz, Th., *The Myth of Mental Illness.* New York, Hoeber-Harper, 1961.
117. Szent-Györgyi, A., In search of new biological dimensions, Perspectives Biol. Med., 4:393–402, 1961.
118. ——, On the possible role of quantum phenomena in normal and abnormal mental function, Publ. Assoc. Res. Nerv. Ment. Dis., 40, 1960.
119. Thompson, R. H. S., and J. N. Cumings, Diseases of the nervous system, in Thompson, R. H. S., and E. J. King, eds., *Biochemical Disorders in Human Disease,* pp. 401–44. New York, Academic Press, 1957.
120. Thudichum, J. L. W., *A Treatise on the Chemical Constitution of the Brain.* London, Balliere, Tindall and Cox, 1884.
121. Tilney, F., and L. S. Kubie, Behavior in the relation to the

development of the brain, Bull. Neurol. Inst., New York, 1:229–313, 1931.
122. Von Bonin, G., and P. Bailey, *The Pattern of the Primate Isocortex*, Bk. II, Pt. 2 (Primatologia). Basel, Karger, 1962.
123. Visscher, M. B., ed., *Methods in Medical Research*, Vol. 4. Chicago, Year Book Publ., 1951.
124. Waelsch, H., ed., *Biochemistry of the Developing Nervous System*, Proc. 1st Int. Neurochem. Symp. London, Oxford, 1954, and New York, Academic Press, 1955.
125. ——, *Ultrastructure and Cellular Chemistry of Neural Tissue*. New York, Hoeber-Harper, 1957.
126. Wald, G., Biochemical evolution, in Barron, E. S. G., ed., *Modern Trends in Physiology and Biochemistry*, pp. 337–77. New York, Academic Press, 1952.
127. Weiss, P., ed., *Genetic Neurology*. Chicago, Univ. of Chicago Press, 1950.
128. ——, Nervous system (neurogenesis), in Willier, B. H., P. Weiss, and V. Hamburger, eds., *Analysis of Development*, pp. 346–401. Philadelphia, Saunders, 1955.
129. Whitelock, O., ed., Symposium: amine oxidase inhibitors, Ann. New York Acad. Sci., 80:551–1045, 1959.
130. Windle, W. F., *Biology of Neuroglia*. Springfield, Thomas, 1958.
131. ——, *New Research Techniques of Neuroanatomy*. Springfield, Thomas, 1957.
132. Wiener, N., *Cybernetics, or Control and Communication in the Animal and Machine,* 2d ed. New York, Wiley, 1961.
133. Wolfle, D., ed., *Symposium on Basic Research*. Washington, D. C., Am. Assoc. Adv. Science, 1959.
134. Wolpe, J., *Psychotherapy; by Reciprocal Inhibition*. Palo Alto, Stanford Univ. Press, 1958.
135. Wolstenholme, G. E., and C. O'Connor, eds., *The Neurological Basis of Behavior,* CIBA Foundation Symposium. Boston, Little, Brown, 1958.
136. Wortis, J., ed., *Recent Advances in Biological Psychiatry*, Vol. 3. New York, Grune and Stratton, 1961.
137. Wyckoff, R. W. G., *World of the Electron Microscope*. New Haven, Yale Univ. Press, 1958.

JOHN H. GADDUM

Chemical Transmission in the Central Nervous System

THE STUDY OF drug effects on the central nervous system has become very popular in recent years, partly because it is of great practical importance to find new drugs for the cure of the mentally ill. The major tranquilizers, such as chlorpromazine and reserpine, have had a profound effect not only on treatment, but also on the attitude of the whole world to mental disease. There are drugs which will relieve anxiety and drugs which will cure depression. Progress in this field has not always been logical; the pharmacologist has provided the tools in the way of new drugs and the clinician has found unexpected uses for them; chlorpromazine was introduced to lower the body temperature and it has been found useful in the treatment of schizophrenia; iproniazid was introduced to cure tuberculosis and has been found useful in the treatment of depression.

The central effects of these drugs can be studied by observing the gross behavior of man or other animals, and numerous tests have been developed to measure behavior and to test the effects of drugs upon it quantitatively (23,

JOHN H. GADDUM is Director of the Institute of Animal Physiology, Babraham, Cambridge, England.

46). This is a complex field, and it is perhaps fortunate for all of us that there is no need for me to say any more about it since it comes under the heading of psychopharmacology, and I have been asked to talk about neuropharmacology. The difference between these two sciences is not clear-cut, but presumably the neuropharmacologist is primarily concerned with the action of drugs on nerves and synapses, and only secondarily interested in the behavior of the whole animal.

THE ACTIONS OF DRUGS ON SYNAPSES
GENERAL CONSIDERATIONS

It is generally agreed that the local release of chemicals plays an important role in the CNS and that some of the mechanisms discovered by the study of chemical transmission in peripheral structures are also present in central structures. In voluntary muscle the end plate is sensitive to drugs and not to electric stimulation. Acetylcholine causes depolarization in the end plate and, when this depolarization is sufficient, an all-or-none impulse is started in the muscle. There is a similar separation of function in the central nervous system (38), where the dendrites are depolarized by chemicals but not by electric potentials; the initiation of impulses in the axon depends on depolarization in the dendrites.

When a substance released by a nerve causes sufficient depolarization to start an impulse, the substance is said to act as a chemical transmitter. It has been suggested that some substances act as modulators, conditioning the CNS without starting impulses. This classification of substances stimulates thought but cannot be rigidly applied, since chemical transmitters may themselves act as modulators.

The release of chemical transmitters appears to be a quantal process. This was shown by Katz (48) for the release of acetylcholine in voluntary muscle, but is also true of the release of another substance in crayfish muscle (24) and may well be a common mechanism. Katz found that electrical records from a resting end plate in voluntary muscle showed small electric changes, which were evidently due to small doses of acetylcholine. They were, for example, increased in size by anticholinesterases and blocked by curare. Each of these miniature end-plate potentials must be due to the simultaneous release from the nerve ending of thousands of molecules of acetylcholine. The acetylcholine in the nerve endings is thought to be contained in small structures with a diameter of about $0.05\,\mu$, which can be seen in an electronmicrograph and which are known as synaptic vesicles (21). Each miniature end-plate potential is thought to be due to the rupture of one synaptic vesicle, and the rate of release of the chemical transmitter depends on the rate of rupture of these vesicles. An impulse in the nerve causes an enormous increase in this rate. When a number of impulses follow one another at a suitable interval, the number of quanta released increases with each successive impulse, which causes facilitation. The crayfish muscle is supplied with an inhibitor nerve, as well as an excitor nerve, and stimulation of this nerve may cause presynaptic inhibition in which the quantal rate falls.

These presynaptic events can all be regarded as modulation, but a pharmacologist is better qualified to speak of postsynaptic events. A small dose of a chemical transmitter may cause some depolarization and make the cell temporarily more excitable without starting an impulse; a larger dose, or more prolonged administration, may start a series of impulses whose frequency depends on the depolarization (this is the normal state of many nerve cells); a still larger dose may cause excessive depolarization and make the cell inexcitable.

When a pharmacological receptor is exposed to large or repeated doses of the same substance, it may become specifically densensitized, and conversely when peripheral tissues are rested by denervation, they may be sensitized (27, 66). This effect may well play a part in modulating the central nervous system; the effect of a small dose of a depolarizing transmitter is likely to be a brief rise of excitability, but prolonged exposure leads to what is sometimes called "tachyphylaxis," in which the drug loses its effect.

The excitatory effects associated with depolarization are due to an unspecific increase of the permeability of the membrane; sodium ions flow into the cell and the membrane potential falls. They are sometimes opposed by inhibitory effects, in which the cell membrane becomes specifically permeable to small ions, such as chloride and potassium. This may lead to an increase of the membrane potential (hyperpolarization). A single substance may cause depolarization at one place and hyperpolarization at another. Acetylcholine, for example, causes depolarization at the end plates in voluntary muscle and hyperpolarization in the heart (8, 29). Two substances may have opposing effects on the same cell, presumably by acting on different pharmacological receptors. For example, adrenaline excites heart muscle where it is directly opposed to acetylcholine. In the smooth muscle of the intestine these two drugs are still opposed, but their roles are reversed: acetylcholine causes depolarization and adrenaline causes hyperpolarization. On other tissues they work together; both of them cause contraction of the splenic capsule.

It is clear that modulation may depend on a direct action of natural substances on synapses, but drugs may affect excitability in other ways, and it is possible that the body releases chemicals which act in some of these other ways. For example, drugs may cause the release of chemical transmitters or inhibit this release; they may preserve them from destruction or prevent their action; they may act like local anesthetics on axons, or produce effects by acting on the blood supply, or act on the metabolic processes of the brain in many other ways. There are clearly many mechanisms for controlling the brain apart from the release of substances with a direct action on synapses, and it is possible that some of these mechanisms are important in the normal action of the brain.

If the action of drugs on the central nervous system is studied by injecting them intravenously, the effects are likely to be indirect. Drugs are, therefore, given by close arterial injection and the effect is localized by destroying the sensory nerves to other tissues supplied by the same blood vessels (7). Drugs given thus may fail to act because of the blood-brain barrier; any effects observed may be due either to an action on nerve cells or to vasomotor effects. It is, therefore, desirable to record the oxygen tension in the tissue by means of an oxygen electrode (11). Vasoconstriction causes a fall of oxygen tension and may affect the electrical activity of nerve cells. It is clearly desirable to distinguish such effects from those caused by the direct action of drugs, and an oxygen electrode makes this possible.

Drugs may be applied to the surface of the tissue in solution or on filter paper, or they may be directly injected into the tissue (49) or into the cerebrospinal fluid (32, 76). The neatest method is by iontophoretic injection with a microelectrode. This technique was first described by Nastuk (57) and was developed by Del Castillo and Katz (20). Concentrated solutions of drugs are contained in microelectrodes and applied to the tissue by passing a current down the electrode which carries the ionized drug into the tissue. In one type of apparatus, five such microelectrodes are formed by fusing five glass tubes together and pulling them out simultaneously. The center tube is filled with saline and the other tubes are filled with solutions of four drugs which can thus be applied to the tissue within 1 or 2 μ of the recording electrode. With this apparatus it is possible to record the effects of drugs and their antagonists on the electric response of a single cell. This cell can then be identified by the methods used in neurophysiology. Its anatomical position is defined with a stereotaxic instrument and its physiological connections may be studied by stimulating the cell itself, and the cells which act upon it, antidromically, and recording the latent periods of the responses.

The effects of drugs on the central nervous system may be detected by recording the movements of muscles, or some other peripheral effect. I am still so old-fashioned as to see some advantages in this approach, but it will never be possible to work out the details in this way; and rapid progress is being made by electrophysiological methods and by the study of extracts of the CNS.

The action of a chemical transmitter may be studied by injecting the substance itself, or by injecting specific antagonists for this substance, or drugs which are known to inhibit its destruction or affect its action in other ways.

Antagonism between drugs may be due to a large number of indirect mechanisms, and this is particularly true for effects on the CNS. Experiments with antagonists are unlikely to throw much light on chemical transmission except when the drugs are competing directly for the same receptors, and it is dangerous to assume that this is so without some evidence. It is reasonable to make this assumption when the following criteria are fulfilled (36):

1. The antagonism is specific—the actions of a control drug are not inhibited.
2. The same two drugs act as antagonists at several different sites.
3. Even when the dose of agonist is increased ten times or more, effective doses of antagonist can be found.
4. The formulae of the two drugs have a pharmacodynamic group in common.

It is not always possible to satisfy all these conditions, and experiments with antagonists in the CNS generally depend largely on the second condition; drugs which have already been established as competitive antagonists by experiments on peripheral tissues are assumed to act competitively in the CNS. Interesting results have been obtained but care is needed in their interpretation. It is well known to pharmacologists that almost any drug will antagonize almost any effect provided you use enough of it. It is, therefore, important to see that the effect is produced by small doses. It is probably also important to compare the general properties of the antagonism between a given pair of drugs when tested on different tissues. For example, various derivatives of lysergic acid are specific antagonists of 5HT (serotonin) on peripheral tissues when used in low concentrations and given an hour or more to act. When high concentrations are used for a short time, the antagonism is unspecific and gives no information about receptors. When these compounds are given to the brain in large doses by close arterial injection, their effect is unlikely to be specific. When they are given systemically and allowed an hour to act, the results may be difficult to interpret owing to slow changes in the animal; but van Gelder,* in my laboratory, was able to show that a skin receptor in the rat was densensitized to 5HT an hour after the subcutaneous injection of 0.01 μg per kilo of methylsergide, while the response to acetylcholine was unchanged. This seems to me to be evidence of competitive antagonism.

ACETYLCHOLINE

Release from peripheral nerves

There is clear evidence that acetylcholine is released and acts as a chemical transmitter in the central nervous system, but there is not equally clear evidence about any other substance, though there can be little doubt that other substances are involved. Much work is being done to discover exactly where acetylcholine is released, and where and how it acts. The best evidence of this is direct and quantitative. In peripheral tissues the acetylcholine can be collected and identified by tests which distinguish it from allied substances such as propionyl choline. It can be estimated and compared with the amount effective for local application. In most experiments of this type, the

* Personal communications.

amount detected has appeared inadequate for transmission and there have been many obvious reasons for the loss.

Recently, however, Krnjevic and Mitchell (50) have obtained evidence that, under optimal conditions, the amount of acetylcholine released in the rat's diaphragm by a single shock to a single nerve-ending (10^{-17} mole) is about the same as the amount that causes an end-plate potential when applied iontophoretically to the rat diaphragm. In this case the acetylcholine causes an end-plate potential when applied by the experimenter, and it can be expected to have the same effect when applied by the nerve in the same quantity. The fact that it does so confirms the theory. The evidence that can be obtained with curare, anticholinesterases and botulin is satisfactory but indirect.

Most of the evidence regarding cholinergic transmission in the CNS is indirect. Early work was reviewed by Feldberg (30, 31).

Cholinacetylase

Distribution. Acetylcholine itself and the enzyme which forms it are both found throughout the whole length of cholinergic neurons. Their distributions are similar and have been used as an index of the distribution of cholinergic nerves. Cholinacetylase forms acetylcholine by transferring the acetyl group from acetyl-coenzyme A to choline. By incubating extracts of tissues with choline under standard conditions the acetylcholine formed can be estimated by its action on the rectus muscle of a frog.

Feldberg and Vogt (33) studied the distribution of this enzyme in the central nervous system and found high concentrations in the anterior roots which consist mainly of cholinergic fibers. They also found high concentrations in some parts of the central nervous system, such as the caudate nucleus and the thalamus, and none at all in other parts, such as the pyramidal tracts and the optic nerves. The distribution of the enzyme is similar to that of acetylcholine, but it can be followed more precisely. Hebb and Silver (40) made a similar survey, with precautions to ensure that there was always sufficient acetyl-coenzyme A in the reaction mixture. With many tissues they obtained very much larger amounts of acetylcholine so that their test was much more sensitive, but, even so, they found negligible amounts in dorsal spinal roots and in optic nerves. Some tissues contained at least 500 times as much of this enzyme as others, and it is reasonable to conclude that these estimates provide an indication of the distribution of cholinergic fibers in the central nervous system.

Centrifugation. The intracellular distribution of cholinacetylase can be followed by homogenizing the tissue and separating various fractions by centrifugation. When a guinea pig's brain is homogenized and centrifuged, acetylcholine and cholinacetylase are both found in the fraction which contains mitochondria but, when this fraction is subdivided by centrifuging in a gradient of sucrose solutions, the acetylcholine and the cholinacetylase can be separated from the mitochondria. The fraction which contains them appears to consist largely of pinched off nerve endings containing synaptic vesicles within which the acetylcholine is thought to be. This fraction also contains high concentrations of various other pharmacologically active substances.

These experiments with homogenates of brain have

given results of great interest, but they tell us nothing about the intracellular distribution of cholinacetylase.

Degeneration of nerves. In experiments with a dog's sciatic nerve, Hebb and Waites (42) found that cholinacetylase was in the microsomal fraction. The fact that cholinacetylase is in this particular fraction and not in free solution is of importance in connection with experiments on cutting nerves. In 1850 Waller (72) showed that when a nerve is cut the peripheral part degenerates and must therefore be normally maintained through the neuron. It is thought that there is a constant flow of substances down the axon of the nerve. When a peripheral nerve is constricted, the diameter of the axons increases in the central stump and decreases in the peripheral part of the nerve (73, 74).

Hebb and her colleagues (41, 42) have studied this flow of substances in the axon by estimating cholinacetylase in peripheral nerves after section. When a dog's sciatic nerve is crushed, cholinacetylase disappears completely from the nerve fibers distal to the crush and accumulates in the peripheral 3 cm. of the central stump during the first few days. Similar studies had previously been made by Sawyer (63) using acetylcholine esterase. After nerve section, this enzyme also accumulates in the central stump, but it only partially disappears from the peripheral part of the nerve. Hebb believes that this is because it is partly outside the axon. During regeneration the amount of the cholinacetylase in the central stump falls as the amount in the peripheral part rises again. After eight weeks the amount in the central part of the nerves falls below normal values as the peripheral part fills up.

It might be thought that these changes were due to

diffusion of the enzyme down the axon, but the microsomes in which it is contained appear to be formed by the disintegration of the endoplasmic reticulum during homogenization. This is a continuous structure and the flow of the enzyme down the nerve can thus hardly be described as diffusion, but may be due to growth. It is suggested that the endoplasmic reticulum is constantly being formed in the nerve cell and pushed down the axon with the cholinacetylase attached to it.

Release

Direct evidence of the release of acetylcholine in the CNS has been hard to get. The amounts detected have been small and the evidence identifying the active substance is incomplete, but in this discussion it will be assumed for simplicity that acetylcholine-like activity was always due to acetylcholine. In 1937 Chang and other colleagues demonstrated the release of acetylcholine into the cerebrospinal fluid (12). Bülbring and Burn (6) perfused eserinized Ringer's solution for a short time through the blood vessels in a dog's spinal cord and showed that stimulation of a sensory nerve caused the release of acetylcholine. They point out, however, that this may have come from voluntary muscle which was also perfused.

MacIntosh and Oborin (54) and Perry (61) detected the diffusion of acetylcholine from the surface of the cerebrum into eserinized Ringer contained in small plastic cups. Angelucci (2) superfused a frog's spinal cord with eserinized Ringer and found that stimulation of the skin released acetylcholine. Mitchell and Szerb (56) have been using plastic cups and also push-pull cannulae (35). In this apparatus Ringer's solution runs down a fine needle into a tissue and is collected by a plastic tube surrounding the needle; in this manner a small area of tissue is constantly washed and active substances are detected in the washings.

In these ways it has been shown that acetylcholine is constantly released from the cerebral cortex at a rate which is directly related to the electric activity. The rate of release is increased by systemic injections of atropine or of the convulsant drug pentylenetetrazol. It is increased by local stimulation of the cortex or by transcallosal stimulation. It is also increased by stimulating the contralateral forepaw. The maximum release was obtained with slow rates of stimulation (about 1/sec.) and at frequencies of 50-100/sec. there was no increase in the rate of release. In some experiments the cannula was in the middle of the caudate nucleus and the rate of release of acetylcholine was increased by stimulation of the anterior part of the motor area at low rates, but not by stimulation of the limbs. It is clear that interesting results may be obtained by these techniques. Other pharmacologically active substances besides acetylcholine may be present in these solutions but, so far, we have failed to show that the rate of release of any substance besides acetylcholine was increased by stimulating nerves.

Action of ACh and Antagonists

The action of acetylcholine in the CNS has been studied by injecting the drug itself, or by injecting other drugs which are known to modify its action. For example, Marrazzi (55) found that transcallosal condition was blocked by atropine and increased by anticholinesterases and came to the reasonable conclusion that it was cholinergic. Bradley and Elkes (4) used similar methods and obtained evidence of a cholinergic mechanism which aroused the electroencephalogram without arousing the cat. Grundfest and his colleagues (38, 39) have used these methods widely.

The neatest way of studying the interaction of drugs on synapses in the CNS is by iontophoretic injection, to which reference has already been made. Using this technique, Eccles and his colleagues (16, 25) have made a careful study of the pharmacology of the spinal cord. The only cells which responded to acetylcholine (cholinoceptive cells) were the Renshaw cells which are normally excited by collaterals of motoneurons and which themselves inhibit motoneurons and so provide negative feedback. This action of acetylcholine is inhibited by the curare-like drug dihydroerythroidine.

Krnjevic and Phillis (51) have been using the same techniques to study the cat's cerebral cortex. Cholinoceptive cells seem to be more common in the brain than they are in the spinal cord. About 15 percent of all the cells tested gave a response to acetylcholine. These cells tended to occur in clusters and often fired in bursts corresponding to the slow wave on the electrocorticogram. They were widely distributed over the primary, somatosensory, visual and auditory receiving areas. The motor area also contains cholinoceptive cells which respond to antidromic stimulation of the pyramids and may be Betz cells.

The anticholinesterase drugs, neostigmine and edrophonium, potentiated the action of acetylcholine, but drugs which specifically antagonize acetylcholine in peripheral tissues were surprisingly ineffective as antagonists in the cerebrum. Atropine and hyoscine had an unspecific depressant action. The drug which antagonized acetylcholine on Renshaw cells (dihydroerythroidine) was ineffective in the cortex. Gallamine is the best antagonist found so far. Perhaps some of the Russian cholinonegative drugs which act on conditioned reflexes (3) will be more effective.

MONOAMINES

Catecholamines

It is convenient to consider four monoamines together—adrenaline, noradrenaline, its precursor dopamine (dihydroxyphenylethylamine) and 5 hydroxytryptamine (5HT, serotonin). They are formed by decarboxylation and their formation is inhibited by α-methyldopa. They are all destroyed by monoamine oxidase and the amount present in the brain may be increased by injecting drugs which inhibit this enzyme, especially if the appropriate precursor is given at the same time. They are all depleted by reserpine which appears to destroy their binding sites in the body, and their peripheral actions are antagonized by ergot alkaloids and dibenyline. The evidence of their physiological importance in the CNS depends, for the most part, on estimations of their concentrations in tissue extracts. There is no evidence that they are released by any particular nerve.

It was the work of Vogt (69) which provided the first convincing evidence that noradrenaline might play a physiological role in the CNS. She found that the noradrenaline in extracts was associated with 5–15 percent of adrenaline and, since no conditions are known which will alter this ratio, the two substances may be considered together. The noradrenaline did not come from vasomotor nerves since it did not disappear when these degenerated. It was concentrated in regions associated with efferent sympathetic pathways and, when these were stimulated by drugs, the concentration fell. In the periphery, noradrenaline is found in nonmedullated postsynaptic neurons, and one theory is that noradrenaline in the CNS is in similar adrenergic neurons. This view is supported by Chruschiel's finding (13) that in homogenates of brain which have been separated by centrifugation, noradrenaline is in the same fraction as acetylcholine, the fraction which appears to consist largely of nerve endings. The arousal caused by activation of the ascending part of the reticular formation is one of the responses which might be due to the release of noradrenaline from central adrenergic nerves, but there is no direct evidence of this.

It is difficult to interpret the evidence of the effect of injections of adrenaline and noradrenaline on the CNS. They cause arousal when given intravenously, but not when given by close arterial injection. When injected into the cerebrospinal fluid they cause sedation. The most likely explanation of this difference is that the arousal caused by intravenous injection is due to the stimulation of peripheral receptors and that the direct effect is sedation, but there is controversy about this. When catecholamines are injected they probably have opposed actions at various sites; some of the effects may be secondary to vasomotor changes (62, 68). Some of these sources of doubt may be eliminated by confining the observations to one synapse or by injecting the catecholamines locally and recording the response at the same place. For example, Trzebski (67) injected catecholamines into the reticular formation and observed increased local electrical activity after a long latent period.

It has been suggested that depletion of noradrenaline causes sedation, but the facts are against this theory. It is true that reserpine causes depletion and sedation; but morphine in the cat causes depletion and excietment, insulin causes depletion without either sedation or excitement and chlorpromazine causes sedation without depletion.

On the other hand, there is better evidence relating increased concentrations of noradrenaline to arousal (65). The therapeutic effect of monoamine oxidase inhibitors in depression is thought to be due to inhibition of the destruction of noradrenaline (which increases the stores and should increase the amount released) but, if this is so, it is surprising that the direct effect of injections of catecholamines appears to be sedation. It is reasonable to conclude that the effects of catecholamines liberated in the brain by nerves may be different from the effects of catecholamines injected by the experimenter, but it would be wrong to suppose that this explanation conceals the fact that we do not really understand what is happening.

Dopamine (dihydroxyphenylethylamine) is the precursor from which noradrenaline is formed in the body, but there is reason to believe that it is also important in its own right. Its distribution in the brain is different from that of noradrenaline, since the highest concentrations are found in the corpus striatum (9). It has little effect on plain muscle.

5 Hydroxytryptamine

The story of the work that has been done on the importance of 5 hydroxytryptamine (5HT, serotonin) in the brain is complex (34, 59). Its distribution resembles that of noradrenaline (1) except that large amounts are present in the caudate nucleus and in the limbic system (58). In centrifuged homogenates the largest amounts are present in the same fraction as acetylcholine, which consists largely of nerve endings (75).

According to Brodie and associates (5), the sedative effect of reserpine is due to the release of 5HT from the stores normally present in the brain. They suggest that it is the released 5HT which depresses the brain, and there are many facts which support this theory. On the other hand, Smith (64) has found that α-methyl-dopa causes a fall of brain 5HT by inhibiting its formation, and this fall is accompanied by sedation. In this case the free 5HT will presumably fall, and he therefore suggests that the sedation due to α-methyl-dopa and that due to reserpine are both due to depletion of the stores of 5HT. This is consistent with the work of Himwich and Costa (44), who raised the concentration of 5HT in the brains of dogs by injecting its precursor, 5 hydroxytryptophan, and observed a curious form of excitement preceded by a brief period of drowsiness. According to Giarman and Schanberg (37), the characteristic effect of a drug is related to its effect on the ratio of the amount of 5HT bound in particles to that free in the supernatant fluid. Various sedative drugs were found to increase the free proportion and stimulant drugs had the opposite effect.

Generally speaking, the accumulation of noradrenaline tends to be accompanied by excitement, and depletion of of 5HT is accompanied by sedation, but the details of the mechanisms are obscure and undoubtedly complicated. LSD is a powerful and specific antagonist of 5HT on peripheral tissues, but not in the cat's brain, where these two drugs each stimulate some centers and inhibit others, so that they are synergists in some places and antagonists in others (70). Various other allied indole compounds have also been found in animal tissues, including tryptamine (43), 3 hydroxytryptamine (10), 6 hydroxytryptamine (47) and melatonin (52).

AMINO ACIDS

The actions of amino acids have been clarified by work in Canberra (17, 19).

When the anion of aspartic acid or glutamic acid is applied by iontophoresis to a neuron in the CNS, it causes depolarization and initiates impulses. Various other amino acids containing two carboxyl groups have the same action. This effect seems to occur in all types of cell in the spinal cord and the cerebral cortex; it occurs in cholinoceptive and in noncholinoceptive cells. It has been used to locate cells which are subsequently tested for their response to other drugs contained in one of the other barrels of the electrode.

These stimulant effects are directly antagonized by the inhibitory amino acids, β alanine and γ aminobutyric acid (GABA), which can be formed from aspartic acid and glutamic acid by α-decarboxylation; and there is reason to believe that this antagonism is competitive (18).

These inhibitory amino acids have been much studied during the last ten years and found to depress a variety of tissues including the stretch receptor in the crayfish (26). It is not possible to discuss all this evidence now. These amino acids are present in very high concentrations in the CNS and must play an important role there. It seems likely that they act as modulators rather than as chemical transmitters. It would be interesting to know more about the factors which control the conversion of glutamic acid into GABA and vice versa.

POLYPEPTIDES

The best known pharmacologically active polypeptides in tissue extracts are oxytocin, vasopressin and substance P. Other active polypeptides, known as kinins, are formed by the action of enzymes on plasma (53), but these are not normally present in tissue extracts and little is known about their possible significance in the CNS.

Vasopressin is present in extracts of the hypothalamus and there is evidence that it is present in granules which pass down nerves to the posterior pituitary whence it is liberated into the blood stream; this process is called neurosecretion.

Substance P is another polypeptide found in extracts of intestine and brain, which is detected by its action on plain muscle and whose function is not known. The following facts suggest that it may have some function, but there is no direct evidence that substance P is liberated in the CNS or that it would have any action if it was.

It has a characteristic distribution in the CNS, some tissues having one hundred times as much as others. The highest concentrations are in gray matter, but it is also present in some nerve tracts. When a nerve to a rabbit's ear was cut and allowed to degenerate, the concentration of substance P in the central stump increased six times in four or five days, while the concentrations in the peripheral part fell to an average of 36 percent in seven days. In these respects substance P behaved like acetylcholinesterase (45). When homogenates of the CNS were centrifuged, substance P was in the same fraction as acetylcholine, cholineacetylase, 5HT and noradrenaline, i.e., the fraction containing nerve endings. None was detected in the mitochondria (14).

Other active substances found in tissue extracts (28, 60) include various fatty acids (71), histamine, ATP and the cerebellar factor (15). There is no convincing evidence that these substances control the brain.

REFERENCES

1. Amin, A. H., T. B. B. Crawford, and J. H. Gaddum, The distribution of substance P and 5-hydroxytryptamine in the central nervous system of the dog, J. Physiol., 126:596–618, 1954.
2. Angelucci, L., Experiments with perfused frog's spinal cord, Brit. J. Pharmacol., 11:161–70, 1956.
3. Anichkov, S .V., Highlights of Soviet pharmacology, Ann. Rev. Pharmacol., 1:21–29, 1961.
4. Bradley, P. B., and J. Elkes, The distribution of cholinergic and noncholinergic receptors in the brain: electrophysiological evidence, in Richter, D., ed., *Metabolism of the Nervous System,* Proc. 2d Int. Neurochem. Symp., Denmark (1956), pp. 515–522. London, Pergamon, 1957.
5. Brodie, B. B., K. F. Finger, F. B. Orlans, G. P. Quinn, and F. Sulser, Evidence that the tranquilizing action of reserpine is associated with change in brain serotonin and not in brain norepinephrine, J. Pharmacol. Exper. Therap., 129:250–56, 1960.
6. Bülbring, E., and J. H. Burn, Observations bearing on synaptic transmission by acetylcholine in the spinal cord, J. Physiol., 100:337–68, 1941.
7. ——, and C. R. Skoglund, The action of acetylcholine and adrenaline on flexor and extensor movements evoked by stimulation of the descending motor tracts, J. Physiol., 107:289–99, 1948.
8. Burgen, A. S. V., and K. G. Terroux, On the negative inotropic effect in the cat's auricle, J. Physiol., 120:449–64, 1953.
9. Carlsson, A., The occurrence, distribution, and physiological role of catecholamines in the nervous system, Pharmacol. Rev., 11:490–93, 1959.

10. ——, M. Lindqvist, T. Magnusson, and B. Waldeck, On the presence of 3-hydroxytryptamine in brain, Science, 127:471, 1958.
11. Carter, D. B., I. A. Silver, and G. M. Wilson, Apparatus and technique for the quantitative measurement of oxygen tension in living tissues, Proc. Roy. Soc. (London) B., 151:256–76, 1959.
12. Chang, H.-C., W.-M. Hsieh, T.-H. Li, and R. K. S. Lim, Humoral transmission of nerve impulses at central synapses, IV and V, Chinese J. Physiol., 13:153–56; 173–86, 1938.
13. Chruschiel, T. L., Observations on the localization of noradrenaline in homogenates of dog's hypothalamus, in *Adrenergic Mechanisms*, pp. 539–43. London, Churchill, 1960.
14. Cleugh, J., and V. P. Whittaker (to be published).
15. Crossland, J., Chemical transmission in the central nervous system, J. Pharm. and Pharmacol., 12:1–36, 1960.
16. Curtis, D. R., and R. M. Eccles, The excitation of Renshaw cells by pharmacological agents applied electrophoretically, J. Physiol., 141:435–45, 1958.
17. ——, J. W. Phillis, and J. C. Watkins, The chemical excitation of spinal neurones by certain acidic amino acids, J. Physiol., 150:656–82, 1960.
18. ——, Actions of amino-acids on the isolated hemisected spinal cord of the toad, Brit. J. Pharmacol., 16:262–83, 1961.
19. Curtis, D. R., and J. C. Watkins, The excitation and depression of spinal neurones by structurally related amino acids, J. Neurochem., 6:117–41, 1960.
20. Del Castillo, J., and B. Katz, On the localization of acetylcholine receptors, J. Physiol., 128:157–81, 1955.
21. De Robertis, E. D. P., and H. S. Bennett, Submicroscopic vesicular component in the synapse, Federation Proc., 13:35, 1954.
22. ——, Some features of the submicroscopic morphology of synapses in frog and earthworm, J. Biophys. Biochem. Cytol., 1:47–58, 1955.
23. Dews, P. B., Analysis of effects of psychopharmacological agents in behavioral terms, Federation Proc., 17:1024–30, 1958.
24. Dudel, J., and S. W. Kuffler, The quantal nature of transmission and spontaneous miniature potentials at the crayfish neuromuscular junction, J. Physiol., 155:514–30, 1961.
25. Eccles, J., The nature of central inhibition, Proc. Roy. Soc. (London) B., 153:445–76, 1961.

26. Elliott, K. A. C., and H. H. Jasper, Gamma-aminobutyric acid, Physiol. Revue, 39:383–406, 1959.
27. Emmelin, N., Supersensitivity following "pharmacological" denervation, Pharmacol. Rev., 13:17–37, 1961.
28. Erspamer, V., Pharmacologically active substances of mammalian origin, Ann. Rev. Pharmacol., 1:175–218, 1961.
29. Fatt, P., The electromotive action of acetylcholine at the motor end-plate, J. Physiol., 111:408–22, 1950.
30. Feldberg, W., Present views on the mode of action of acetylcholine in the central nervous system, Physiol. Rev., 25:596–642, 1945.
31. ——, Acetylcholine, in Richter, D., ed., *Metabolism of the Nervous System*, Proc. 2d Int. Neurochem. Symp., Denmark (1956), pp. 493–510. London, Pergamon, 1956.
32. ——, and S. L. Sherwood, Injections of drugs into the lateral ventricle of the cat, J. Physiol., 123:148–67, 1954.
33. ——, and M. Vogt, Acetylcholine synthesis in different regions of the central nervous system, J. Physiol., 107:372–81, 1948.
34. Gaddum, J. H., Recent work on 5-hydroxytryptamine and lysergic acid derivatives, Proc. XX Intern. Physiol. Congr. Abstr. Review, 442–55, 1956.
35. ——, Push-pull cannulae, J. Physiol., 155:1–2P, 1961.
36. ——, Antagonisms between drugs, Neuro-Psychopharmacol., 2:19–24, 1961.
37. Giarman, N. J., and S. M. Schanberg, Drug-induced alterations in the intra-cellular distribution of 5-hydroxytryptamine in rat's brain, Biochem. Pharmacol., 8: 6, 1961.
38. Grundfest, H., General problems of drug actions on bioelectric phenomena, Ann. New York Acad. Sci., 66:537–91, 1957.
39. ——, The interpretation of electrocortical potentials, Ann. New York Acad. Sci., 92:877–89, 1961.
40. Hebb, C. O., and A. Silver, Choline acetylase in the central nervous system of man and some other mammals, J. Physiol., 134:718–28, 1956.
41. ——, Gradient of choline acetylase activity, Nature, 189:123–25, 1961.
42. Hebb, C. O., and G. M. H. Waites, Choline acetylase in antero- and retro-grade degeneration of a cholinergic nerve, J. Physiol., 132:667–71, 1956.
43. Hess, S., B. G. Redfield, and S. Udenfriend, Tryptamine

in animal tissues following administration of iproniazid, Federation Proc., 18:402, 1959.
44. Himwich, W. A., and E. Costa, Behavioral changes associated with changes in concentrations of brain serotonin, Federation Proc., 19:838–45, 1960.
45. Holton, P., Substance P concentration in degenerating nerve, in Schachter, M., ed., *Polypeptides which Affect Smooth Muscle and Blood Vessels*, pp. 192–96. London, Pergamon, 1960.
46. Jacobsen, E., The comparative pharmacology of some psychotropic drugs, Bull. World Health Organization, 21:411–93, 1959.
47. Jepson, J. B., S. Udenfriend, and P. Zaltzmann, The enzymic conversion of tryptamine to 6-hydroxy-tryptamine, Federation Proc., 18:254, 1959.
48. Katz, B., Microphysiology of the neuro-muscular junction. A physiological 'quantum of action' at the myoneural junction, Bull. Johns Hopkins Hosp., 102:275–312, 1958.
49. Kennard, D. W., Local application of substances to the spinal cord, in Malcolm, J. L., and J. A. B. Gray, eds., *The Spinal Cord*, CIBA Foundation Symposium, pp. 214–22. London, Churchill, 1953.
50. Krnjevic, K., and J. F. Mitchell, The release of acetylcholine in the isolated rat diaphragm, J. Physiol., 155:246–62, 1961.
51. ——, and J. W. Phillis, Sensitivity of cortical neurones to acetylcholine, Experientia, 469, 1961.
52. Lerner, A. B., and J. D. Case, Melatonin, Federation Proc., 19, (1):590–92, 1961.
53. Lewis, G. P., Active polypeptides derived from plasma proteins, Physiol. Rev., 40:647–76, 1960.
54. MacIntosh, F. C., and P. E. Oborin, Release of acetylcholine from intact cerebral cortex, Proc. XIX Intern. Physiol. Congr. Abstr. Commun., pp. 580–81, 1953.
55. Marrazzi, A. S., Some indications of cerebral humoral mechanisms, Science, 118:367–70, 1953.
56. Mitchell, J. F., and J. C. Szerb (to be published).
57. Nastuk, W. L., Membrane potential changes at a single muscle end-plate produced by transitory application of acetylcholine with an electrically controlled microjet, Federation Proc., 12:102, 1953.

58. Paasonen, M. K., P. D. MacLean, and N. J. Giarman, 5-hydroxytryptamine (serotonin, enteramine) content of structures of the limbic system, J. Neurochem., 1:326–33, 1957.
59. Page, I. H., Serotonin (5-hydroxytryptamine); the last four years, Physiol. Rev., 38:277–35, 1958.
60. Paton, W. D. M., Central and synaptic transmission in the nervous system (pharmacological agents), Ann. Rev. Physiol., 20:431–70, 1958.
61. Perry, W. L. M., Central and synaptic transmission, Ann. Rev. Physiol., 18:279–308, 1956.
62. Rothballer, A. B., The effects of catecholamines on the central nervous system, Pharmacol. Rev., 11:494–547, 1959.
63. Sawyer, C. H., Cholinesterases in degenerating and regenerating peripheral nerves, Am. J. Physiol., 146:246–53, 1946.
64. Smith. S. E., The pharmacological actions of 3, 4-dihydroxyphenyl-α-methylalamine (α-methyldopa), an inhibitor of 5-hydroxytryptophan decarboxylase, Brit. J. Pharmacol., 15:319–27, 1960.
65. Spector, S., P. A. Shore, and B. B. Brodie, Biochemical and pharmacological effects of the monoamine oxidase inhibitors iproniazid, 1-phenyl-2-hydrazinopropane (JB 516) and 1-phenyl-3-hydrazinobutane (JB 835), J. Pharmacol. Exp. Therap., 128:15–21, 1960.
66. Thesleff, S., Effects of motor innervation on the chemical sensitivity of skeletal muscle, Physiol. Rev., 40:734–52, 1960.
67. Trzebski, A., The action of adrenaline, noradrenaline and monoamine oxidase inhibitors injected directly into the reticular formation of the brain stem by means of a micro-injection method, Acta Physiol. Pol., 11:905–07, 1960.
68. Vane, J. R., *Adrenergic Mechanisms.* London, Churchill, 1960.
69. Vogt, M., The concentration of sympathin in different parts of the central nervous system under normal conditions and after the administration of drugs, J. Physiol., 123:451–81, 1954.
70. ——, C. G. Gunn, and C. H. Sawyer, Electroencephalographic effects of intraventricular 5HT and LSD in the cat, Neurology, 7:559–66, 1957.
71. Vogt, W., Naturally occurring lipid-soluble acids of pharmacological interest, Pharmacol. Rev., 10:407–35, 1958.
72. Waller, A. V., Experiments on the section of the glossopharyn-

geal and hypoglossal nerves of the frog and observations of the alterations produced thereby in the structure of their primitive fibres, Phil. Trans. Roy. Soc., London (B), 140:423–29, 1850.
73. Weiss, P., Protoplasm synthesis and substance transfer in neurons, Proc. XVII Intern. Physiol. Congr., p. 101, 1947.
74. ——, and H. B. Hiscoe, Experiments on the mechanisms of nerve growth, J. Exp. Zool., 107:315–95, 1948.
75. Whittaker, V. P., The isolation and characterization of acetylcholine containing particles from brain, Biochem. J., 72:694–706, 1959.
76. Winterstein, H., The actions of substances introduced into the cerebrospinal fluid and the problem of intracranial receptors, Pharmacol. Rev., 13:71–107, 1961.

GEOFFREY W. HARRIS

The Development of Neuroendocrinology

IT IS KNOWN that the central nervous system regulates the activity of the neurohypophysis by means of the supraopticohypophysial nerve tract and controls the activity of the adrenal medulla by means of the splanchnic nerves and sympathetic chains. However, the greater number of the ductless glands—the thyroid, adrenal cortex, ovaries and testes—are brought under neural control through the intermediation of the anterior pituitary gland. Brain-anterior pituitary-target gland relationships therefore form a key link in the neuroendocrine hierarchy.

The pituitary gland as an endocrine organ

The first views on brain-pituitary relationships were those expressed by the Greeks. They believed that the blood in the left ventricle of the heart was mixed with air in this chamber and imbued with vital spirits. This blood, after transport to the brain, was thought to be endowed with animal spirits, the waste products of the reaction being passed by the funnel-shaped infundibulum to the pitu-

GEOFFREY W. HARRIS is Chairman of the Department of Neuroendocrinology, Maudsley Hospital Institute of Psychiatry, University of London, England.

itary gland and thence by ducts to the nasal cavity. Here the excreta formed the pituita or nasal mucus.

These ideas were generally adopted for the next fourteen hundred years. From Galen (130–201) to Vesalius (1514–1564) little change was made in the theory that the pituitary acts as a type of cerebral kidney. Vesalius in *De Humani Corporis Fabrica* accepts the Galenic doctrine that various channels in the third ventricle, and elsewhere, lead phlegm into the infundibulum and pituitary gland, but disagrees with Galen's view that the sphenoid is perforated by foramina in a sponge-like manner beneath the pituitary. He states (Singer, 110), "The phlegm descends to the palate through these foramina which pass down thereto, and largely through the foramen of the second pair of nerves [this would be the oculomotor] near the eye-sockets. It goes thence through a spacious fissure [superior orbital fissure] and through many other foramina, unknown to other professors of dissection, to the cavity of the nose." He believed that normally no phlegm decends to the nose through the olfactory bulbs but suggested that perhaps this does happen when the brain is afflicted by excess of phlegm, when there is "an itching, heat, pain and inability to smell, and symptoms of that kind, which we experience daily in catarrhs at the top of the nose."

The view that mucus is transmitted to the nasal cavity by the pituitary was denied by two workers in the seventeenth century, Conrad Victor Schneider of Wittenburg (1614–1680) and Richard Lower (1631–1691) (Fulton, 34). Schneider, in 1655, showed that the cribriform plate of the ethmoid transmitted only the olfactory nerve and its branches and that fluid from the cerebrum did not pass to the nose through the foramina in this bone. In 1657, in his six-volume work *De Catarrhis*, Schneider argued that the nasal secretions in catarrh are derived from the nasal mucosa and not from the brain. Richard Lower was also instrumental in overthrowing the classic doctrine. In his *Dissertatio de Origin Catarrhi* in 1670, Lower describes experiments in which he removed the calvarium and injected water into the jugular veins. He observed that although fluid appeared in the ventricles and infundibulum and the subdural space was distended, no fluid passed into the pharynx or nares. Lower also described the case of a young boy with hydrocephalus, in which no loss of fluid had been observed from the nose even though the pituitary showed signs of compression. From this and other evidence he concluded that the nasal mucus was formed in situ by the nasal mucosa and urged the abolition of therapies which aimed at purging the brain for infections of the upper respiratory tract. Lower's writings are also significant in that he seems, at this early date, to have had some conception of internal secretion. He writes, "Whatever serum is separated into the ventricles of the brain and issues out of them through the infundibulum to the glandula pituitaria, distills not upon the palate but is poured again into the blood and mixed with it." As Rolleston (95) says, it is tempting to read into this statement the germ of the concept of internal secretion made rather more than a century before de Bordeu's speculations (1775) which have usually been regarded as the first glimmerings of present knowledge.

Between the time of Lower and the beginning of the twentieth century, little advance was made regarding the possible functions of the pituitary gland, although, in

1724, Santorini described the two lobes of the gland. In 1778 van Soemmering proposed the alternative name "hypophysis" and in 1838 Rathke elucidated the embryology of the pituitary. Views as to the function of the organ were, however, varied. Hypotheses were advanced that the gland controlled the blood flow to the brain or, according to Magendie (1842), played a role in regulating the pathway of the cerebrospinal fluid. With the advent of Darwin's views on evolution, the general theory became prominent that the gland was a vestigial relic and Macalister, in 1889, expressed his opinion that the pituitary was "probably the rudiment of an archaic sense organ." Even in 1909 Murray's *New English Dictionary on Historical Principles* stated that the pituitary was a "small bilobed body of unknown function attached to the infundibulum at the base of the brain, originally supposed to secrete the mucus of the nose."

During the nineteenth century, however, ideas of internal secretion were being considered. The first clear evidence for endocrine function as understood today came from the work of Berthold of Gottingen in 1849 (11). He published a four-page paper in which he described the effects of castration and testicular transplantation in cocks. In this account he gave very clear and precise data that the testes produced an internal secretion responsible for the male character and instincts of the cock. Berthold did in fact emphasize, in one part of his paper, a facet of neuroendocrinology that has only received serious attention in the last few years. He said, "the consensus in question is conditioned by the secretion of the testes, i.e., by their action on the blood on the whole organism of which, it must be admitted, the nervous system forms a very substantial part."

Thus he emphasized the importance of the feed-back action of hormones on the brain, that is, the way in which the behavior of an animal is adjusted to its endocrine balance. Some fifty years later, studies of the effects of administration of posterior pituitary extracts by Oliver and Schäfer (87), Howell (65), Dale (19), Ott and Scott (88) and others drew attention to the possibility that the neurohypophysis was an endocrine gland. Several workers in California—Drs. H. M. Evans, P. E. Smith and B. M. Allen—then played a leading role in showing the importance of the adenohypophysis as a gland of internal secretion. They demonstrated clearly the deficiency phenomena produced by hypophysectomy in amphibians and mammals, the repair of these deficits by pituitary implants or by administration of crude extracts, and were able to produce experimental hyperpituitarism by injection of anterior pituitary extracts. Since that time Dr. Evans and his school of workers in Berkeley have reported classic studies in which they purified and analyzed the actions of the greater number of the anterior pituitary hormones.

It is now possible to state that the anterior pituitary gland secretes a number of hormones which regulate body growth and metabolism (growth hormone), mammary function (lactogenic hormone), and also maintain and regulate the activity of other endocrine organs, such as the ovary and testis (follicle-stimulating hormone, luteinizing hormone), the thyroid (thyrotrophic hormone) and the adrenal cortex (adrenocorticotrophic hormone).

The central nervous system and endocrine function

The first indication that the brain might exert an influence over anterior pituitary function was derived from the observation that certain sensory, environmental stimuli may markedly affect the function of the target glands under pituitary control. A clear example of this may be seen in the effects of environmental stimuli on ovarian function. Such data have been accumulating for many years. I am grateful to Dr. J. D. Green for drawing my attention to what are probably the earliest recorded observations in this field; first, those of Martin Lister in 1675 (73) who, in a letter to John Ray, wrote that swallows may be induced to lay more eggs than usual by removing an egg daily from the nest; and second, those of Haighton (48), who described the ovarian changes in rabbits which are reflexly induced by coitus. In 1905 Heape (62) gave a detailed description of reflex ovulation in the rabbit, and later workers showed that the stimulus of coitus acted by certain nervous reflex paths to excite the pituitary secretion of gonadotrophic hormone and thus cause ovulation. Dr. F. H. A. Marshall of Christ's College Cambridge studied the effect of various environmental stimuli on reproductive rhythms; and, in 1936 and 1942, he published two reviews (77, 78) that were of major importance in focusing attention on the role of exteroceptive factors in influencing gonadal function. Marshall (78) summarized his views by saying, "it would appear certain that many external factors which regulate the cycle act through the intermediation of the central nervous system upon the anterior pituitary, this gland playing the part of a liaison organ between the nervous system which is affected by stimuli from without and the endocrine system."

The idea that an endocrine gland may be excited to increased activity by a whole variety of traumatic, physical and emotional stimuli—by what may therefore be called "nonspecific stimuli" of a noxious or stressful type—seems to have occurred to two workers and to both in respect to the adrenal glands. Walter Cannon (16), who studied mainly the effects of different emotions on bodily systems and structures, summarized in 1915 his findings regarding the autonomic and adrenal medullary responses to "emergencies." The response of the body to a whole variety of stimuli, calling for a reaction of the organism often in terms of fight or flight, results in central nervous activation of the peripheral autonomic nervous system and thus the adrenal medulla. Hans Selye (108) in 1936 published the first of a long series of articles dealing with a stereotyped response of the body to a whole variety of damaging agents. One aspect of this response was the increased activity and enlargement of the adrenal cortex. It is now clear that stress stimuli act largely through the central nervous system to increase pituitary discharge of adrenocorticotrophin and thus lead to adrenal cortical activation.

Many other examples of the influence the brain exerts on anterior pituitary, and thereby on other glandular, activity could be quoted. The influence of a cold environment in stimulating thyroid activity, the effect of stress stimuli in inhibiting thyroid function, and the effects exerted by suckling or emotional upsets on mammary function are among these. The general picture that emerges, then, is that many different receptor areas of the brain, acting via reflex paths to the hypothalamus, may modify anterior pituitary activity and the activity of its target glands in such a way as to adjust the endocrine balance of the organism in accordance with its environment. Such a scheme is represented diagrammatically in Figure 1.

The neural control of one other endocrine gland, the

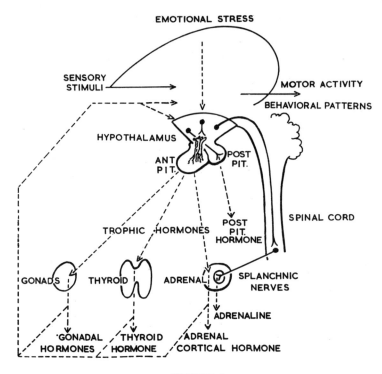

FIGURE 1

The reciprocal relationships existing between the central nervous system and the endocrine system.

neurohypophysis, deserves mention here, though it is not dealt with in detail later. This gland is stimulated to increased secretory activity by sensory stimuli, especially those derived from the genital and mammary regions, by stress stimuli and by hypertonicity of the blood. The classic studies of Fisher, Ingram and Ranson (28), and of Verney (114) have thrown much light on the secretomotor innervation of the neurohypophysis by the supraopticohypophysial tract, on the effects of section of this tract, and on factors regulating neurohypophysial function.

The pituitary stalk as the "final common path" of an effector system

From the beginning, the pituitary stalk has been suggested by many workers to be the anatomical path by which the central nervous system controls the adenohypophysis. The relationship the stalk bears to the pituitary gland may be compared, in many respects, with the relationship of a motor nerve to its muscle. Sherrington's phrase "final common path" was therefore coined (Harris, 54; Sayers and Royce, 100) to denote the final pathway by which a multitude of nervous reflexes affects the activity of the effector organ, in this case the gland.

Twenty to thirty years ago much time was spent in searching for a secretomotor innervation to the pituitary. Nerve fibers were described by some workers as passing into the pars distalis from the pituitary stalk, from the carotid sympathetic plexus and from the petrosal nerves. The more detailed histological studies of Rasmussen (94), Green (40) and Wingstrand (116), however, failed to reveal the presence of a secretomotor nerve supply to the gland cells. In retrospect it seems probable that many of the fibers previously described in the gland as nerve fibers were reticular connective tissue fibers.

The absence of nerve fibers passing from the stalk to the anterior pituitary gland left open again the question of which anatomical structure in the stalk served to transmit central nervous influences over the gland. A few years before this problem became prominent, Popa and Fielding (91, 92) had described the hypophysial portal system—a system of vessels connecting specialized capillaries in the upper end of the pituitary stalk, the region known as the "median eminence" of the tuber cinereum, with the sinusoids in the pars distalis. The anatomical existence of these vessels was soon confirmed by Wislocki and King (119) and by Wislocki (117, 118), but differences of opinion were expressed as to the direction of blood flow in the portal vessels. Perhaps for this reason, and perhaps also from the confusion that arose concerning the nomenclature of these vessels, little attention was paid to the possibility that they might occupy an important key position in neuroendocrine relations. Green and Harris (42) reinvestigated the anatomy of the portal vessels in a variety of mammals and listed evidence indicating that the vessels might form the anatomical path in the pituitary stalk by which the hypothalamus regulated anterior pituitary activity. Shortly afterward, the view of Wislocki that the blood flows from the median eminence to the pituitary gland was substantiated by direct microscopic observation of the vessels in living amphibians (Houssay, Biasotti and Sammartino, 64; Green, 39) and in mammals (Green and Harris, 43; Török, 113; Worthington, 122) and interest began to be focused on the functional significance of this vascular system.

The anatomical relations of the vessels, the established direction of blood flow within them, and their origin in a region of high capillary permeability (Wislocki and King, 119) suggested the possibility that they might act as a transport system to the pituitary gland for humoral agents liberated by hypothalamic nerve fibers or cells, and that such a neurohumoral mechanism might form the means of hypothalamic regulation of the anterior pituitary gland. One objection raised to this hypothesis was that a septum intervenes between the posterior and anterior lobes of the pituitary in some forms, which would presumably interrupt the pathway of a portal vascular system. This objection was shown to be invalid for at least one of these forms, the porpoise (*Phocaena phocaena*), by Harris (52), who found that the portal vessels were carried by the zona and pars tuberalis of the gland around the rostral border of the septum. It then seemed likely that if the vessels did in fact form the anatomical basis for such a fundamental function, they would be widely distributed in all vertebrates. That this is so was found by Green (40), who made a detailed investigation of hypophysial blood supply in many vertebrates. In his important publication Green concludes:

It is a remarkable fact that the hypophysioportal circulation shows such minor variations between related species and that the variations described can be followed with such ease in so orderly a manner in a phylogenetic series. Such constancy suggests a functional significance. Were it not so, wide variations might be expected to occur, since the general morphology of the pituitary is anything but constant.

The first direct evidence that the portal vessels are concerned with anterior pituitary function came from experiments dealing with pituitary stalk section. It was found (Harris, 53) that section of the stalk in rats may be followed by rapid regeneration of the vessels across the site of section. In female rats in which vascular regeneration occurred, estrous cycles returned, postoperatively, in 22 out of 23 animals. Animals in which such regeneration was prevented by placement of a plate between the cut ends of the stalk remained anestrous after operation, and the postmortem examination of these animals revealed atrophy of the reproductive organs. Thus a correlation was found to exist between the presence of the hypophysial portal vessels and the normal release of gonadotrophic hormone.

Similar results were found in the ensuing years by different workers in a number of forms—ducks, mice, ferrets, rabbits, sheep and goats—and indicated by findings in the human. However, owing largely to the technical difficulties inherent in such experiments, a few workers obtained equivocal results which they interpreted differently. For this reason a study was made (Harris and Jacobsohn, 58) of the functional activity of pituitary transplants. Pituitary tissue was placed in the subarachnoid space of hypophysectomized rats, first under the hypothalamus where it became largely vascularized by the portal vessels (Figure 2), and second under the temporal lobe of the brain when it became vascularized by cortical and dural vessels. Hypophysectomized female rats with hypothalamic transplants had regular estrous cycles, mated, became pregnant, delivered living young and showed normal milk secretion. Their thyroid and adrenal glands were well maintained. On the other hand, the control animals with temporal lobe transplants remained anestrous after operation and were found to have gonadal, thyroidal and adrenal atrophy. Thus again a correlation was found to exist between anterior pituitary function and the vascularization of the gland tissue by the portal vessels.

During this work it was also noticed that pituitary tissue taken from rats only a few days old would maintain normal adult function if transplanted under the hypothalamus of an adult hypophysectomized female animal. Further, pituitary tissue taken from male animals would support normal estrous cycles if similarly transplanted into hypophysectomized females. Thus reproductive maturity is not solely dependent on aging of pituitary tissue, and the arhythmic or rhythmic nature of the reproductive

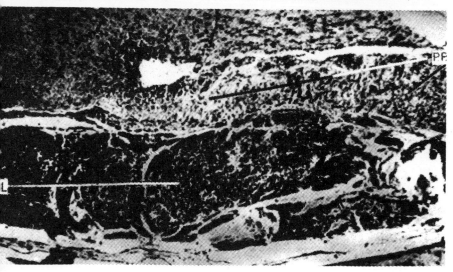

FIGURE 2

crophotographs of sagittal sections through the hypothalamus, a pituitary graft and base of skull of a rat. (× 42.) The upper photo is an unstained section 100 μ thick. The vascular connections pass from the primary plexus (PP) of the hypophysial portal vessels to the graft (AL). dia ink injected specimen). The lower photo is an adjacent section through the same specimen to show the graft consisting of anterior lobe tissue (AL); (PP) is primary plexus. (Haematoxylin and eosin) (From Harris and Jacobsohn, 58) processes normally seen in male or female rats is not due to sexual differentiation of pituitary tissue. It appears that anterior pituitary tissue remains plastic in nature and that the pattern of activity it shows is dependent on the stimuli it receives from the central nervous system via the portal vessels.

These results have been substantiated by subsequent studies. For example, Martinez and Bittner (79) observed that hypophysectomized female mice, bearing either female or male pituitary grafts in the sella turcica, showed complete and normal estrous cycles, and that hypophysectomized male mice, bearing either male or female pituitary grafts in the sella turcica, failed to show cyclic secretion of gonadotrophins. Nikitovitch-Winer and Everett (85) investigated the effects of transplanting pituitary tissue to the renal capsule and then, at a later date, retransplanting the same tissue back under the hypothalamus. From their results they conclude that "the extreme functional deficiencies observed when the pars distalis is removed to sites distant from the hypothalamus are caused by the loss of hypothalamic influences, stimuli which are probably mediated by the hypophysial portal vessels. This is true with respect to secretion of ACTH, TSH, FSH, and LH." Smith (111) studied the effects of transplanting pituitary tissue (obtained from siblings) under the hypothalamus of male rats that had been hypophysectomized two to five months previously. Following the transplantation, a pronounced gain in body weight, activation of the thyroid and a structural repair of the adrenals and reproductive organs occurred. In view of the fact that the pars distalis of the pituitary shows normal activity only when supplied with blood which has passed through the capillaries in the median eminence of the tuber cinereum (with blood, that is, which has come into immediate relationship with the nervous tissue in this structure), the hypothesis was formulated that anterior pituitary function is maintained at a normal level, and adjusted by nervous reflexes according to the needs of the external environment by means of humoral substances liberated from nerve endings in the median eminence and carried to the gland by means of the portal vessels (Taubenhaus and Soskin, 112; Harris, 50; Green and Harris, 42).

In more recent years this idea has received indirect support from facts which have accumulated regarding the mode of secretion of the neurohypophysis. Originating largely from the work of Bargmann (4), and Scharrer and Scharrer (103), it seems likely that the posterior pituitary hormones are not formed by gland cells in the neurohypophysis, but are formed by hypothalamic neurons (of the supraoptic and paraventricular nuclei) which send nerve fibers into the neurohypophysis. The hormones are then liberated from the nerve terminals directly into the blood vessels of the gland and carried via the blood stream to exert their effects upon distant organs. The neurohypophysis, which includes the median eminence as well as the neural lobe, may then be a structure specialized by reason of its vascular permeability for the distribution of hormonal substances released from nerve endings into the blood stream. On the one hand, these substances, vasopressin and oxytocin, are known to be octapeptides carried by the vascular system to exert an action on the breast, uterus, kidney and other structures; and on the other, they may well be polypeptides carried by the portal vessels to exert an action on the anterior pituitary gland. Figure 3

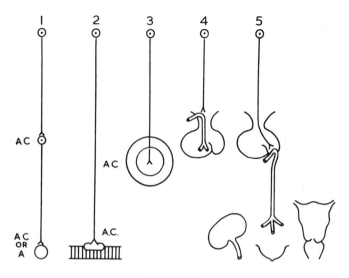

FIGURE 3

Different systems in which humoral transmission of stimuli may occur: 1 autonomic nervous system, 2 neuromuscular ending, and 3 sympathoadrenal medullary ending. In these three systems it is established that a cholinergic (AC) or adrenergic (A) substance is liberated from the nerve terminal and acts directly on the effector cell. 4 hypothalamoadenohypophysial system, in which the evidence indicates that a short vascular pathway intervenes between the nerve terminal and the effector cell situated in the anterior pituitary. 5 Hypothalamoneurohypophysial system, in which a long vascular path (the systemic circulation) intervenes between the nerve terminal in the neural lobe and the effector cells in the kidney, breast, or uterus. (From Harris, 54)

expresses these views diagrammatically and compares them with the well-established findings regarding humoral transmission in the peripheral nervous system.

At least two major lines of enquiry are being pursued at the present time to test the above hypothesis. The first utilizes histological and histochemical techniques to see if any changes occur in the predicted neurosecretory mechanism in the median eminence in correlation with stimuli known to evoke changes in anterior pituitary secretion.

Several studies have been made on birds (Benoit and Assenmacher, 10; Oksche, Laws, Kamemoto and Farner, 86; Kobayashi and Farner, 71), in which histologically observable neurosecretory material is more abundant in the median eminence than in mammals. Although the results are not entirely clear-cut, the studies indicate that there may be a diminution in the amount of neurosecretory material in the median eminence, as opposed to the neural lobe, simultaneous with an increased release of gonadotrophic hormone. Osmotic stimuli, on the other hand, seem to exert an effect primarily on the neurosecretory content of the neural lobe. Also from the cytological point of view, it will be of great interest to see what electron-microscopic techniques reveal regarding neurovascular relationship and the structure of nerve terminals in the median eminence under varying conditions of pituitary stimulation. Several important studies of the type have been made on the neurohypophysis (Green and Van Breeman, 45; Duncan, 24; Fujita, 33; Palay, 89; De Robertis, 21), but the attention was largely focused on the transportation and release of posterior pituitary hormones. Green and Maxwell (44), for example, describe several types of granules in the nerve fibers of the hypothalamo-hypophysial tract. It would be important to know whether microsomal granules occur in the nerve terminals in relation to the primary plexus of the hypophysial portal vessels, and whether any change in these granules is detectable on nervous reflex activation of the anterior pituitary. Correlated studies of median eminence fractions obtained by differential centrifugation over a density gradient might also be enlightening. A good review by Hebb (63) deals with such studies made on nervous tissue in general.

A second line of approach to the problem of humoral control of anterior pituitary function has involved the study of the effects of hypothalamic or neurohypophysial extracts, and other substances of physiological interest, on pars distalis function. The work of Taubenhaus and Soskin (112) and that of Markee, Sawyer and Hollinshead (76) was influential in drawing attention to this approach. Although many studies have now been made of humoral stimulation of the release of ACTH, TSH, and LH from the pituitary, and many suggestions have been made concerning the chemical nature of humoral transmitter substances transplanted by the portal vessels, the final data regarding the chemical structure and the identification in portal vessel blood of any one of these substances remain to be found. Saffran, Schally, Guillemin and their coworkers have, however, produced a most interesting series of papers over the last six years dealing with the properties of a corticotrophin-releasing factor (CRF) obtained from hypothalamic and neurohypophysial tissue (Saffran, Schally and Benfey, 97; Guillemin, Hearn, Cheek and Housholder, 46; Guillemin and Schally, 47; Schally and Guillemin, 102). These workers have obtained a peptide similar to, but not identical with, vasopressin that is potent in releasing ACTH from anterior pituitary tissue both in vivo and in vitro. This material has been obtained in a highly purified form (Schally, Andersen, Lipscomb, Long and Guillemin, 101). It will be of great interest, once its chemical structure is known, to see whether it is present in portal vessel blood in amounts proportionate to the rate of secretion of ACTH.

Quite recently some definite progress has been made with regard to humoral stimulation of LH release. Campbell, Feuer, Garcia and Harris (Harris, 57; Campbell et al., 14) have obtained a crude extract of median eminence tissue of rabbits, cattle and monkeys that is effective in evoking LH secretion and subsequent ovulation when infused into the anterior pituitary gland of rabbits by microcannulae. These same extracts were tested in the same laboratory by Nikitovitch-Winer on rats and found to be active in causing ovulation in the nembutal-blocked rat. Almost simultaneous with these publications are the findings of McCann and Taleisnik (81, 82) from America and of Guillemin (personal communication) from Paris, that they have obtained median eminence and hypothalamic extracts also active in evoking LH release in rats. Once the nature and identity of this luteinizing-releasing factor (LRF) has been established, measurement of its concentration in portal vessel blood under different conditions of reflex activation of the pars distalis will be necessary. This latter would seem to be a simpler project in the case of LRF than CRF, since nonspecific stimuli involved in the collection of the blood will probably influence the release of LRF less than CRF.

The hypothalamus and the anterior pituitary gland

During the last fifteen years many studies have been devoted to problems concerned with hypothalamic and anterior pituitary relationships. A majority of these have utilized the time-honored methods in neurological research of making lesions in, or electrically stimulating, different areas of the hypothalamus and observing the effects produced on the activity of the anterior pituitary target glands. A precise assessment of hypothalamic control over the pars distalis still awaits the development of simple methods which may be used in a routine manner for measuring the concentration of the six trophic hormones in arterial and in pituitary venous blood. Various methods for the assay of these hormones in body fluids are now available but they tend to be too complex, laborious and insensitive for general use. Perhaps the future preparation of pure pituitary hormones from different species will allow the development of immunological methods of assay as a satisfactory procedure.

The CNS tends, in general terms, to control the activity of somatic structures by a balance of both excitatory and inhibitory influences. The influence of the hypothalamus over anterior pituitary function seems to afford another example of such control. Since disconnection of the gland from the brain, by severing the pituitary stalk, results in loss or diminution of the release of the hormones (the exception being the luteotrophic hormone, Everett, 27) it would seem that the main balance of the hypothalamus over pituitary secretion is excitatory.

The most satisfactory techniques for studying the effects of electrical stimulation of the hypothalamus are those that can be used without the complication of anesthesia, such as the remote control method in which the stimulus is induced into an implanted coil and electrode unit, or one involving chronically implanted electrodes with leads exteriorized through the skin. By such methods it has been found that stimulation of various hypothalamic areas may lead to increased discharge of luteinizing hormone (LH), adrenocorticotrophic hormone (ACTH), and thyrotrophic hormone (TSH). The release of LH, as assessed by consequent ovulation in the rabbit, has been reported by Harris (49, 51), Haterius and Derbyshire (61), Markee, Sawyer and Hollinshead (75) and many others to follow stimulation of the tuberal and more posterior hypothalamic areas. Similar results have since been obtained in cats and rats whose spontaneous ovulation has been blocked. The secretion of ACTH induced by hypothalamic stimulation was first reported using the lymphopenic response in the rabbit (de Groot and Harris, 20) and the eosinopenic response in the dog (Hume and Wittenstein, 66). Similar results, using more direct criteria of adrenal cortical activation such as the concentration of adrenal steriods in systemic or adrenal venous blood, have since been obtained in a variety of forms, including primates (Mason, 80). The areas involved in ACTH regulation appear to be the anterior median eminence, posterior tuberal and mammillary regions. More anterior regions of the hypothalamus, however, have been implicated in the control of TSH secretion. Harris and Woods (60) found that remote control stimulation of the supraoptic area in rabbits evoked a manifold increase in thyroid activity, as measured by the release of thyroidal I^{131} and the blood level of PBI^{131}. Confirmation of this result was obtained in anesthetized rabbits (Campbell, George and Harris, 15) in which the content of labeled hormone in thyroidal venous blood was studied.

Lesions have been utilized in many studies to investigate the role of the hypothalamus in regulating anterior pituitary activity, and results confirming those described previously have been obtained. Thus, appropriately placed lesions will block the release of LH and the ovulation that normally follows coitus in certain forms; will prevent the stress-induced release of ACTH; and have been found to result in marked reduction of TSH secretion and thyroid activity, and abolition of the TSH discharge normally evoked by environmental cold. From the conclusions of stimulation experiments, these results would have been expected. More surprising, however, was the finding that hypothalamic lesions could induce release of follicle-stimulating hormone (FSH) under conditions in which it would not otherwise have occurred. This was first observed by Donovan and van der Werff ten Bosch (22). These workers placed lesions in the anterior hypothalamus of female ferrets during the winter anestrum, and found that such animals developed estrus in mid-January instead of early March as was the case with the controls. Repetition of the study yielded essentially similar results, although lesions placed in the posterior hypothalamus, thalamus or amygdala were without effect. Possible explanations of these results were that the lesions caused compression and therefore stimulation of surrounding hypothalamic areas, or that damage removed the effects of an inhibitory zone.

Closely analogous to these results are the cases of precocious puberty which occur in young children in association with hypothalamic tumors. In these cases hypothalamic injury results in increased discharge of FSH in the prepubertal condition. In both the ferrets and these cases of pubertas praecox, the hypothalamic lesion results in the transformation of a quiescent reproductive state, anestrous or prepubertal, to that of an active condition.

In view of such considerations, Donovan and van der Werff ten Bosch (23) studied the effect of similar hypothalamic lesions on the reproductive organs of immature rats. They found that lesioned animals reached puberty approximately a week before normal litter mates or blank operated controls. These studies have been confirmed, in the main, by Bogdanove and Schoen (12) and by Gilliot and Ganong (see Ganong, 35). The most likely explanation of these results is that the ovary of the immature or anestrous female is secreting a certain basal level of steroid hormone which is sufficient to act on the hypothalamus to depress FSH liberation by the pituitary. If the anterior hypothalamus is damaged, this feed-back mechanism is blocked and FSH discharge increases with a resultant activated state of the reproductive organs.

It has long been known that the hormones secreted by the pituitary target glands exert a feed-back action on some central structure to influence the secretion of the pituitary trophic hormones. Thus they form part of a servo-mechanism by which their own release is regulated. Flerkó of Hungary has recently performed a signal service by attracting attention to this aspect of neuroendocrine relations. Although the level, hypothalamic or pituitary, at which the feed-back action occurs is not yet clear, some facts have been established. Investigations have followed three main lines: First, the effect of hypothalamic lesions in blocking the action of administered exogenous hormones has been studied. Flerkó (29) found that lesions in the region of the hypothalamic paraventricular nuclei prevented the gonadal atrophy induced by estrogen administration, and Flerkó and Illei (30) found that similar lesions interfere with the inhibition of gonadotrophic secretion produced by testosterone propionate. On the other hand, hypothalamic lesions do not interfere with the depression in TSH release after thyroxine administration, or the diminution of ACTH secretion which follows injection of adrenal steroids. Second, the effects of hypothalamic lesions in blocking the effects of reduced endogenous hormone levels in blood has been investigated. It has in general been observed that such lesions prevent the compensatory increase in trophic hormone secretion that normally follows unilateral adrenalectomy or thyroidectomy. Third, the local application of target organ hormones to different regions of the hypothalamus or pituitary has been accomplished by transplant (Flerkó and Szentágothai, 31) or microinfusion techniques (Von Euler and Holmgren, 115). It has been found that estrogen-sensitive structures are present in the hypothalamus and thyroxine-sensitive in the pituitary gland itself. From these results and others, it is usually concluded that the feed-back of ovarian hormones in regulating gonadotrophin release occurs at a hypothalamic level, but that the feed-back of thyroxine may occur at a pituitary level. On the other hand, even though the pituitary cells are sensitive to thyroxine, it may well be that future work will show the presence of hypothalamic chemoreceptors sensitive to this hormone, and at a lower threshold than that of the pituitary.

At the present time (1961) a dim picture of hypothalamic-anterior pituitary relationships begins to emerge, though many important questions remain unanswered. It is particularly intriguing that certain of these would appear soluble by presently available techniques and that, in many cases, pioneer studies have already been performed which can form a basis for future work. Four such fields of enquiry, which may be classified as anatomical, physiological, electrophysiological and biochemical, may be mentioned.

1. The anatomy of the fiber tracts and nuclear connections of the hypothalamus, and of reflex pathways entering the hypothalamus, are little known. There are no answers

to such apparently simple questions as: "By what nervous pathways is the coital stimulus transferred to the anterior pituitary gland in the rabbit?" Several recent papers may point the way to a better understanding in this field. The extensive study by Nauta and Kuypers (84) of the projection from the brain stem reticular formation to the hypothalamus may be mentioned in this respect. There is however, as yet, very little anatomical substrate for the experimental observations made on hypothalamic-endocrine functions.

2. Physiological observations on the function of the spinal cord in relation to motor control, on the brain stem respiratory mechanism and other central nervous functions, have usually started from basic information in which the nervous structure involved has been studied in isolation. Thus, a knowledge of the reflex properties of the isolated spinal cord formed a basis for the integrative functions of "higher" neural regions. In this light, what is urgently required is knowledge of the neuroendocrine mechanism of an animal with the hypothalamus isolated from the rest of the central nervous system but still in normal relationship with the pituitary gland. Such studies must necessarily be made in chronic preparations. It might then be possible to get a clearer idea of the integrative effects of limbic and other higher nervous structures in mediating emotional and other reflex endocrine changes. The studies of Bard and Macht (3) and Woods and Bard (120, 121) are of especial interest in this connection.

3. Electrophysiological studies of the hypothalamic responses to hormonal administration, or to stimuli known to evoke changes in anterior pituitary secretion, offer an important approach to problems concerning localization of

hypothalamic function and the neural mechanisms involved. In recent years Sawyer and Kawakami (99) have published accounts of their experiments dealing with the effect of hormones on the thresholds of electrical stimulation of the EEG arousal and the EEG afterreaction phenomena. Also working in Los Angeles, Cross and Green (18) studied the electrical changes occurring in the neurons of the nucleus supraopticus following administration of hypertonic saline, which is known to evoke secretion of the antidiuretic hormone.

4. Biochemical studies and work involving the uptake of radioactive hormones by the hypothalamus are just beginning. It is known for example, that isotopically labeled thyroidal, ovarian and adrenal cortical hormones seem to be selectively localized in various areas of the hypothalamo-pituitary complex, but whether this local concentration of hormone is related to the feed-back mechanism regulating pituitary secretion, or to phenomena related to behavioral responses, is unknown. The physicochemical changes occurring in hypothalamic neurons under the influence of estrogens, which are responsible for the dramatic behavioral changes seen in estrus, are yet an unexplained field.

Extra-hypothalamic regions of the central nervous system and anterior pituitary function

Gloor (37) states, "On the basis of comparative anatomical studies, it becomes therefore evident that it is mainly the ring of old cortex surrounding the hilus of the hemisphere and the old striatum, which represent the telencephalic formations closely connected with the hypothalamus." These formations comprise the hippocampus; the cortex of the piriform lobe; the orbitofrontal, anterior insular and cingular cortex, the amygdala and globus pallidus. These structures, which are richly interconnected by the cingulum and fasciculus uncinatus, would seem to act as intermediary between the neocortex and neostriatum and the hypothalamus. They are richly connected with the hypothalamus via the fornix, the stria terminalis, direct amygdalo-hypothalamic connections, and the orbito- and pallido-hypothalamic system of fibers.

That the hippocampus and amygdala might be implicated in this control of pituitary secretion was first suggested by the important work of Klüver and Bucy (70), Klüver and Bartelmez (69) and Klüver (68). Klüver and Bartelmez (69) reported that bilateral removal of both prefrontal and temporal lobes in a monkey was followed by hyperplasia of the rete ovarii and an extensive endometriosis; results which the authors felt might be attributable to degeneration of the fibers to the hypothalamus. Electrical stimulation of the amygdala has been found to result in ovulation in rabbits (Koikegami, Yamada and Usui, 72), cats (Shealy and Peele, 109) and rats in which spontaneous ovulation has been blocked by continuous illumination (Bunn and Everett, 13). Porter (93) drew attention to the possibility that the hippocampal area, in particular the uncus, might be related to ACTH secretion. Other workers have found that electrical stimulation of the amygdala in the monkey (Mason, 80) and in different animals (Endröczi, Lissák, Bohus and Kovács, 26) excites ACTH release, although Yamada and Greer (123) observed that bilateral destruction of the amygdala in the male rat does not interfere with the secretion of either TSH or ACTH. Stimulation of the hippocampus has, on the other hand, been found to exert a depressant effect on ACTH release (26, 80), and Mason (80) found that bilateral section of the fornix in the monkey obliterated the normal diurnal variation which occurs in adrenal cortical activity.

The importance of the brain stem reticular formation in so many central nervous processes has naturally raised the possibility that this structure might play a role in regulating endocrine function. Various workers have studied the effect of placing lesions in (Critchlow, 17), or sectioning (Egdahl, 25; Anderson et al., 1; Royce and Sayers, 96), the midbrain and obtained evidence that alterations in the cyclic release of gonadotrophic hormone, or in the basal or stress-induced release of ACTH, do in fact occur. The mechanisms involved are far from clear, though Critchlow (17) has implicated a definite fiber tract, the mammillary peduncle, as being involved in the ovulatory mechanism in the rat.

In attempting to summarize the part played by extra-hypothalamic regions of the central nervous system in regulating endocrine function, the views of Gloor (37) may be quoted:

The hypothesis is advanced that the limbic system does not fundamentally integrate the functions it is capable of influencing by its activity, but rather acts as a modulator of functional patterns integrated at the level of the hypothalamus and brain stem tegmentum. It may represent an important link between neocortex and hypothalamo-tegmental formations conveying information enabling the subcortically induced activities to adapt to the patterns organized by the neocortex. Such adaptational mechanisms may be involved in emotional expression. It is finally suggested that the limbic system may take part in activation of the correlated endocrine mechanisms.

Pharmacological blockade of the neuroendocrine axis

A rapidly expanding frontier of investigation concerns the pharmacological field in which search is made for drugs to block the neuroendocrine axis at different points. From the clinical point of view, it is more useful to have drugs which can block rather than stimulate glandular activity, since a deficiency of hormone can be more easily rectified by administration of hormonal extracts or synthetic material. Certain blocking drugs have been known for some time, e.g., alloxan and its action on the pancreas; but systematic study has only recently started in what promises to be an increasing and fruitful field of research.

It would be possible here to discuss many different aspects of this subject, for example, the action of drugs on the secretion of gonadotrophic hormones, originating from the important publications of Markee, Everett and Sawyer (74). However discussion will be limited to the drugs affecting the neuro-hypophysial-adrenocortical axis. Reference may be made to the excellent and comprehensive review by Gaunt, Chart and Renzi (36) for a more detailed account. Figure 4 illustrates the central nervous-hypophysial-adrenal axis and shows points along this axis at which drugs, available at the present time, may block the chain of events.

A. *Neural level.* The central nervous system normally acts to maintain a certain basal secretion of ACTH, and to excite an increase in ACTH release in response to a whole variety of environmental stimuli, classified as noxious in type and therefore as stresses. On administration of certain central nervous system depressants, an initial increased secretion of ACTH is followed by a reduction of secretion and a reduction in the response to stress. Some of these drugs are morphine, dibenzyline, atrophine, bar-

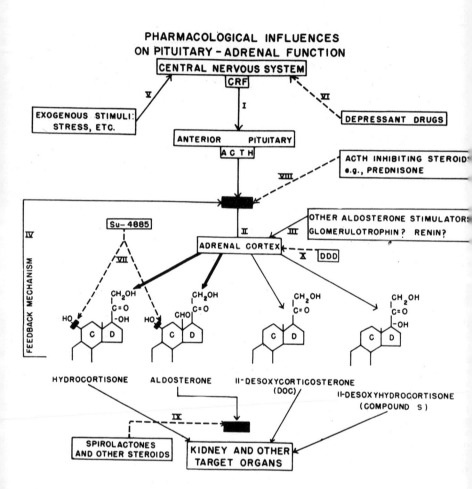

FIGURE 4

Some aspects of normal pituitary-adrenal function and the site of action of drugs which modify it. Physiological factors are illustrated by solid arrows, pharmacological ones by broken arrows. Inhibition or blockade is indicated by black boxes.

Physiological: In the central nervous system a corticotrophin releasing factor (CRF) stimulates (I) the anterior pituitary to secrete corticotrophin (ACTH), which in turn causes (II) the secretion of the several adrenal hormones (only four of which are shown). Other humoral agents also influence (III) aldosterone secretion, but their identity is uncertain. Of the corticoids illustrated, hydrocortisone and aldosterone are the most important; the other two being produced normally in insignificant amounts. Hydrocortisone serves (IV) as the main natural inhibitor of ACTH secretion, thus providing a self-limiting regulatory system. It may act directly on the pituitary, the hypothalamus, or both. Exogenous stimuli (V) can, however, stimulate ACTH secretion, thus overriding the feed-back mechanism (IV). Not illustrated is the possibility, for which some evidence exists, that the CNS exerts inhibitory as well as stimulating influences on ACTH secretion.

Pharmacological: Depressant drugs, analgesics, tranquilizers, and so on, can inhibit (VI) the secretion of ACTH by either direct effects on the nervous system or by insulating it from exogenous sensory stimuli (V). Su-4885 acts (VII) primarily to inhibit the enzymatic hydroxylation of adrenal hormones at carbon-11, thus preventing or reducing the secretion of hydrocortisone and aldosterone. The lack of hydrocortisone eliminates the feed-back mechanism (IV), permitting excess secretion of ACTH. Under these circumstances an excess of ACTH causes an out-pouring of 11-deoxycorticosterone and 11-deoxyhydrocortisone. These are mineralocorticoids and their levels may rise sufficiently to cause sodium retention. If a steroid which inhibits ACTH (VIII) is given with Su-4885, all corticoid secretion is reduced, and sodium diuresis and reduction of edema may occur. The spirolactones act (IX) as competitive inhibitors of aldosterone and block its effect on the kidney and at other sites. This results in sodium diuresis and a compensatory hypersecretion of aldosterone. The DDD-like compounds cause (X) cellular atrophy of the adrenal, lack of responsiveness to ACTH, and decreased secretion probably of all corticoids. (From Gaunt, Chart, and Renzie, 36) biturates, ether, reserpine and chlorpromazine. There is evidence that certain of these drugs act primarily on the reticular activating system to prevent reflex nervous stimuli from reaching the hypothalamus.

B. *Feed-back action.* Steroids which have a feed-back action to inhibit ACTH secretion, like the natural glucocorticoids, and yet which have comparatively slight action on the adrenal cortical target organs, are effective in bringing about a diminution in adrenal cortical activity. It is not known whether the main site of action of steroidal feed-back to inhibit ACTH release is exerted at a neural or hypophysial level.

C. *Adrenal cortical level.* The insecticide DDD and related compounds cause cellular atrophy of the adrenal cortex and render it unresponsive to ACTH. Other drugs, such as SU-4885, act at an adrenal level by blocking 11β-hydroxylating mechanisms and thus the formation of aldosterone, hydrocortisone and corticosterone.

D. *Target-organ level.* The spirolactones act at a kidney level to prevent the mineralocorticoids from exerting their sodium-retaining effects and may thus be of value as diuretics.

Although many of the drugs mentioned are too toxic, or the doses required are too great to be of therapeutic use, the results so far obtained are of great value in that research for less toxic related compounds has been greatly stimulated, and the analysis of the mode of action of the present drugs throws much light on the basic mechanisms involved.

The central nervous feed-back action of hormones

In the account given above attention has been devoted to the regulation exerted by the central nervous system over the anterior pituitary gland and its target glands. The CNS is intimately concerned with the maintenance of a normal level of endocrine activity and also with the adjustment of endocrine function according to the needs of a varying environment.

The reciprocal side of the above picture concerns the effects exerted on the brain by varying concentrations of target gland hormones in the blood. On the one hand, this is dealing with the influence of the CNS on endocrine activity, and on the other with the influence of hormones on brain activity and behavior. This latter aspect of neuroendocrine relationships has received much less attention than the former, though studies during the last few years have shown that a wide and very important field awaits attention. Although it is clear that changes in the blood concentration of, for example, gonadal and thyroidal hormones, may profoundly change the behavior of an animal, only fragments of information are available as to the extent these hormones cross the blood-brain barrier, as to which nuclear groups or neural mechanisms they act upon and whether they are locally concentrated in different brain regions or not, and as to the biochemical and biophysical processes they influence.

The actions of hormones on the development of the central nervous system. It is now being recognized that important relationships between the endocrine and nervous systems exist in fetal and immature animals. The work of Jost (67) has done much to elucidate the degree of endocrine activity existent in the fetus. There can be little doubt that most target-glands are actively secreting in the later stages of fetal life and that to some extent their activity is pituitary dependent and regulated by a feed-back mechanism. The onset of full adult activity in the gonads occurs dramatically, and this event has long been distinguished by the term "puberty." What is not so clearly appreciated is that a type of "puberty" probably occurs in the case of the thyroid and adrenal cortex also. By puberty, in this latter case, is understood the onset of full adult thyroidal and adrenal cortical function and the capacity of regulating these functions, probably by nervous reflex action, according to environmental needs. It is likely that both for the gonads and the other glands one of the factors regulating the onset of puberty is maturation of the central nervous system, which is itself in part dependent on the blood concentration of hormones secreted by the immature glands. The general outline of this picture can be most clearly visualized by considering neural-gonadal relationships in the immature animal.

It was logical that originally the onset of sexual activity should have been attributed to an aging process of gonadal tissue. This idea was found to be false long ago, when Foa (32) showed that the ovaries of immature animals functioned as adult organs if transplanted into mature animals. It was then suggested that the puberty was dependent on aging of anterior pituitary tissue. But again it was shown (Harris and Jacobsohn, 58) that newborn pituitary tissue transplanted under the hypothalamus of hypophysectomized adult animals will maintain adult reproductive functions. It then seemed likely that the CNS is concerned in some way with the onset of full gonadal function, a view compatible with the observation that damage to the hypothalamus, often by hamartomata, in immature children may result in precocious puberty (Bauer, 7). This clinical condition has now been reproduced experimentally by Donovan and van der Werff ten Bosch (23). The fact that hypothalamic lesions are related to precocious puberty suggests that the central nervous system normally acts by inhibiting reproductive activity in the immature animal. Some indirect evidence exists which suggests that the low level of gonadal hormones in the circulation of immature forms acts via the nervous system to inhibit gonadotrophin secretion by the anterior pituitary gland. Hypothalamic lesions would then be effective by blocking this inhibiting feed-back action of ovarian and testicular hormones.

It is probable that gonadal steroid hormones act on the nervous system of immature animals not only to hold in abeyance full anterior pituitary function but also to differentiate some neural mechanism into a male or female type. Adult male mammals show a steady level of reproductive activity, reflecting a constant secretion rate of gonadotrophic hormones. On the other hand, the female shows a rhythmic activity, be it estrous or menstrual, reflecting the alternating levels of FSH and LH release. The arhythmic or rhythmic nature of these processes is probably determined by some central process. It is clear that it cannot be due to the presence or absence of ovaries, since many workers have found that ovaries transplanted into adult castrated males fail to form corpora lutea or show rhythmic activity. Attention was first drawn to the possibility of sexual differentiation of some central structure by the important work of Pfeiffer in 1936 (90). He showed that ovaries transplanted into adult male rats, castrated when adult, do not undergo rhythmic changes but show a constant state of follicular ripening, and that ovaries transplanted into adult male rats, castrated soon after birth, do undergo rhythmic changes of follicular ripening, ovulation and corpora lutea formation. It thus appeared to Pfeiffer that male and female rats are born with some inherent mechanism that would evoke a rhythmic release of gonadotrophins, and that during maturation this mechanism, which he located in the anterior pituitary gland, is differentiated under the action of androgens secreted by the immature testis into an arhythmic male type mechanism. That the rhythmicity or arhythmicity of the adult reproductive processes is not located in the pituitary gland, is clearly shown by the fact that male pituitary tissue, transplanted under the hypothalamus of hypophysectomized female rats, has been found capable of maintaining normal female reproductive functions. Thus any sexual differentiation that occurs in the immature animal, under the influence of testicular secretion, most likely involves the nervous system.

The early work of Pfeiffer has recently been confirmed by more precise experiments involving the injection of male hormone into newborn female rats. The work of Segal and Johnson (107), Barraclough (5), and Barraclough and Gorski (6) deserves special mention in this context. These workers, whose results have now been confirmed and extended by other groups, found that a single subcutaneous injection of testosterone propionate (10-1000 μg) in oil into a five-day-old female rat consistently results in failure of ovulation and a persistent estrous state when the animal becomes adult. The term "persistent estrus" is almost established in the literature to describe the state shown by these animals, but is of doubtful validity. The vaginal epithelium does show a consistently cornified state, reflecting the constant state of ripe follicles in the ovaries, and thus may be truly referred to as a persistently estrous-like vagina. On the other hand these animals will rarely accept the male, and since by definition estrus is the period when the female is receptive, they are less persistently-estrous in the behavioral sense than are normal female rats. Terms which differentiate between an estrous-like state of the reproductive organs and behavioral estrus, are badly needed for precise description.

In order to investigate the anatomical site of action of the postnatally administered steroids, Segal and Johnson (107) transplanted the pituitary glands of anovular or persistent-estrus female rats into the empty sella turcica of hypophysectomized normal female rats and found that these glands were capable of maintaining normal gonadotrophic secretion and maintaining normal female reproductive processes in this environment. Thus the administered testosterone has clearly not acted by differentiating anterior pituitary tissue and the probability exists that it acts upon neural, possibly hypothalamic, mechanisms concerned with the pattern of gonadotrophin release and the pattern of mating behavior.

Using testosterone as a "hormonal dissecting tool," Gorski and Barraclough (38) have found interesting data indicating that two distinct CNS regions regulate ovulation and mating behavior. They gave single injections of 5, 10 or 20 μg of testosterone propionate to groups of female rats at five days of age; 5 μg was found to block ovulation in only a proportion of animals, 10 μg blocked ovulation in all. However rats injected with 10 μg were found to mate copiously when placed with males, although 1.0 mg renders them sexually unreceptive.

A more detailed study of the mating behavior patterns

of female and male rats, injected when immature with testosterone and estrogen, respectively, has been made by Levine and Harris (unpublished). Observations made under standardized conditions showed that the testosterone-treated females were totally unreceptive, and that in response to advances by the male reacted with vigorous back-kicking and often with aggressiveness. Some of these females were ovariectomized and treated with estrogen and progesterone; they still remained unreceptive although the aggressive behavior was diminished. The estrogen-treated males showed no response to an estrous female in 45 percent of cases. The remainder exhibited some mounting responses, and occasionally achieved intromission, though all failed to ejaculate.

The above results may then be generalized to indicate that both male and female rats are born with a female type neural mechanism, so far as reproductive patterns are concerned. Under the influence of the secretion of the immature testis the central nervous system of the prepuberal male rat becomes differentiated so that the pattern of gonadotrophin release by the pituitary becomes differentiated into a male pattern, and sexual behavior becomes differentiated into that of the normal male. Possible application of these results into the field of clinical psychiatry are obvious, viz., that certain groups of male homosexuals may be due to lack of hormonal differentiation of the nervous system in the prepubertal state. Since the threshold for behavioral differentiation seems, in the rat, to be higher than that for the pattern of gonadotrophin release, it would be of interest to see whether a rhythm of gonadotrophin secretion occurs in any type of male homosexual. A further corollary of the above experimental findings would indicate the grave danger of treatment with steroid hormones in the human infant; the possible consequences of such treatment not being apparent for many years.

The action of hormones on the brain and behavioral patterns in the adult. The work of Beach, Bard and others, which has been the subject of important reviews (Beach, 9; Bard, 2), suggests that the normal pattern of sexual behavior shown by subprimate females is dependent on the direct action of ovarian hormones on the central nervous system. The results of experiments involving the placement of lesions indicate that ovarian hormones act to lower the threshold of some "center," situated in the posterior hypothalamus or upper midbrain, so that certain sensory stimuli, normally supplied by the presence of a male, act to evoke a stereotyped pattern of behavior. In the female cat, this extensive pattern, which lends itself well to experimental study, may be analyzed into "courtship," "mating," and "afterreaction" activities.

Investigation of the neural basis of sexual behavior by the placement of nervous lesions, with observation of any loss of the overt reaction, is open to the objection that damage to mesencephalic or diencephalic structures may interfere with many bodily functions that might indirectly affect the complex, integrated pattern of reflexes under study. Stimulation of the behavioral pattern, by localized chemical (hormonal) or electrical stimuli would seem to give more significant results. With this view in mind, Harris and Michael (59) and Harris (56) investigated the effects of implanting minute fragments of solid estrogenic substances, stilbestrol di-n butyrate, in different regions of the brain of ovariectomized cats. The untreated ovariectomized cat shows a consistent loss of sexual receptivity and atrophy of the reproductive tract. If the animal is injected systemically with an estrogen development of the reproductive tract (with cornification of the vaginal epithelium), after a predictable latent period estrous behavior is brought about. It was thought that implantation of minimal amounts of estrogen in the brain might activate the development of behavioral estrus by a local action, which could be detected by observing sexual receptivity in the presence of an atrophic reproductive tract. After preliminary experiments to see which stilbestrol had an appropriate absorption rate, amounts of stilbestrol di-n butyrate varying from 50–800 μg were fused to the tips of stainless steel needles and inserted into the brain by stereotaxic methods. The sexual responsiveness of these animals was tested by a standard routine over long periods. It was found that implants in the cerebellar hemisphere, preoptic region, frontal white matter, caudate nucleus, thalamus and amygdaloid region were not effective in evoking estrous behavior. However stilbestrol implants in the posterior hypothalamus, in the region of the mammillary bodies, resulted in the development of full mating behavior in thirteen of the seventeen animals studied and in many of these the reproductive tract remained in a completely undeveloped state. Thus after mating tests the vaginal smear of these animals showed sperm mingled with non-cornified epithelial cells. From these and other control experiments it was concluded that the stilbestrol implants in the posterior hypothalamus had activated some neural mechanism responsible for the integration of simple reflexes into the total pattern of sexual behavior. Further experiments by Michael (83), in which radioactive stilbestrol butyrate has been implanted in a similar manner shows that the stilbestrol diffuses into a strictly localized area around the implant.

Another view regarding the localization of the sexual behavior "center" in the cat has been put forward by Sawyer (98). He proposes, on the basis of lesioning experiments, that the region concerned is localized in the anterior hypothalamus of the cat, although in the rabbit he believes the region of the mammillary bodies is involved.

Although the evidence is strong that the basic mechanism for the display of sexual responsiveness is localized in the posterior hypothalamus, there can be little doubt that other regions of the brain influence this mechanism by reflex pathways, excited both by conditioned and unconditioned stimuli, in the higher mammals. Thus, removal of cerebral cortical tissue in rats resulted in diminution of sexual activity in proportion to the amount of cortical tissue removed (Beach, 8). Interference in the preoptic hypothalamic area results in a temporary hypersexual state in male and female rats. Attention was first drawn to the influence of the temporal lobe, in this respect, by Klüver and Bucy (70), when they found that bilateral temporal lobectomy resulted in a hypersexual state in monkeys. Subsequent studies by Schreiner and Kling (104, 105, 106) tended to implicate the amygdaloid nuclei in the hypersexual state induced in the cat, lynx, monkey and agouti by temporal lobe lesions. A detailed study by Green, Clemente and De Groot (41) in male cats, however, showed that the effective lesion associated with the hypersexual state is in the pyriform cortex adjacent to the amygdaloid nucleus.

Although attention has been focused here on the effect of sex hormones on the brain and behavior, there is a wide field of investigation awaiting further research on adrenocortical- and thyroid-brain relationships. Doubtless radioactive hormones will play an important part in such research.

It is interesting to note that our present views on the function of the pituitary stalk, that is the structure which is largely responsible for linking the brain and endocrine system into a functioning unit, is in principle a return to the views expressed by the Greeks. It is currently held that a transport of material of neural origin occurs both in the nerve fibers of the hypothalamo-hypophysial tract to the neurohypophysis and in the hypophysial portal vessels to the adenohypophysis.

The future of neuroendocrinology, a subject of study which barely existed thirty years ago and which has grown in this time with remarkable speed, will involve a multidisciplinary approach. The techniques of anatomy, cytology, electron microscopy, physiology, pharmacology, biochemistry (probably involving much use of isotopically-labeled hormones) and experimental psychology are clearly going to be involved in collaborative studies on the many problems facing us at the present time. Such collective work can obviously be best performed in an institute in which the many disciplines involved in brain research are represented. It is therefore with a sense of confidence and anticipation for the future development of neuroendocrinology that one views the opening of the Brain Research Institute here at UCLA.

REFERENCES

1. Anderson, E., R. W. Bates, E. Hawthorne, W. Haymaker, K. Knowlton, D. McK. Rioch, W. T. Spence, and H. Wilson, The effects of midbrain and spinal cord transection on endocrine

and metabolic functions with postulation of a midbrain hypothalamico-pituitary activating system, Recent Prog. in Hormone Research, 13:21–66, 1957.
2. Bard, P., The hypothalamus and sexual behavior, Research Publ. Assoc. Research Nervous Mental Disease, 20:551–79, 1940.
3. Bard, P., and M. B. Macht, The behaviour of chronically decerebrate cats, in Wolstenholme, G. E. W., and C. M. O'Connor, eds., *Neurological Basis of Behaviour*, CIBA Foundation Symposium, pp. 55–7. London, Churchill, 1958.
4. Bargmann, W., Über die neurosekretorische Verknüpfung von Hypothalamus und Neurohypophyse, Z. Zellforsch. Mikroskop. Anat., 34:610–34, 1949.
5. Barraclough, C. A., Production of anovulatory, sterile rats by single injections of testosterone propionate, Endocrinology, 68:62–67, 1961.
6. —, and R. A. Gorski, Evidence that the hypothalamus is responsible for androgen-induced sterility in the female rat, Endocrinology, 68:68–79, 1961.
7. Bauer, H. G., Endocrine and other clinical manifestations of hypothalamic disease, J. Clin. Endocrinol., 14:13–31, 1954.
8. Beach, F. A., Effects of cortical lesions upon the copulatory behaviour of male rats, J. Comp. Psychol., 29:193–245, 1940.
9. —, *Hormones and Behaviour*. New York, Hoeber, 1948.
10. Benoit, J., and I. Assenmacher, Le controle hypothalamique de l'activité préhypophysaire gonadotrope, J. Physiol. Path. Gén., 47:427–567, 1955.
11. Berthold, A. A., Transplanation der Hoden, Arch. Anat. Physiol. Berlin, pp. 42–46, 1849.
12. Bogdanove, E. M., and H. C. Schoen, Precocious sexual development in female rats with hypothalamic lesions, Proc. Soc. Exp. Biol. Med., 100:664–69, 1959.
13. Bunn, J. P., and J. W. Everett, Ovulation in persistent-estrous rats after electrical stimulation of the brain, Proc. Soc. Exp. Biol. Med., 96:369–71, 1957.
14. Campbell, H. J., G. Feuer, J. Garcia, and G. W. Harris, The infusion of brain extracts into the anterior pituitary gland and the secretion of gonadotrophic hormone, J. Physiol., 157:30–31P, 1961.
15. —, R. George, and G. W. Harris, The acute effects of injec-

tion of thyrotrophic hormone or of electrical stimulation of the hypothalamus on thyroid activity, J. Physiol., 152:527–44, 1960.
16. Cannon, W. B., *Bodily Changes in Pain, Hunger, Fear and Rage,* 2d ed. New York, Appleton, 1915.
17. Critchlow, V., Blockade of ovulation in the rat by mesencephalic lesions, Endocrinology, 63:596–610, 1958.
18. Cross, B. A., and J. D. Green, Activity of single neurones in the hypothalamus: effect of osmotic and other stimuli, J. Physiol., 148:554–69, 1959.
19. Dale, H. H., The action of extracts of the pituitary body, Biochem. J., 4:427–47, 1909.
20. De Groot, J., and G. W. Harris, Hypothalamic control of the anterior pituitary gland and blood lymphocytes, J. Physiol., 111:335–46, 1950.
21. De Robertis, E., Morphological bases of synaptic processes and neurosecretion, in Kety, S. S., and J. Elkes, eds., *Regional Neurochemistry,* Proc. 4th Int. Neurochem. Symp., pp. 248–57. London, Pergamon, 1961.
22. Donovan, B. T., and J. J. van der Werff ten Bosch, The relationship of the hypothalamus to oestrus in the ferret, J. Physiol., 147:93–108, 1959.
23. ——, The hypothalamus and sexual maturation in the rat, J. Physiol., 147:78–92, 1959.
24. Duncan, D., An electron microscope study of the neurohypophysis of a bird, *Gallus Domesticus,* Anat. Record, 125:457–71, 1956.
25. Egdahl, R. H., Adrenal cortical and medullary responses to trauma in dogs with isolated pituitaries, Endocrinology, 66:200–16, 1960.
26. Endröczi, E., K. Lissák, B. Bohus, and S. Kovács, The inhibitory influence of archicortical structures on pituitary-adrenal function, Acta Physiol. Hung., 16:17–22, 1959.
27. Everett, J. W., Functional corpora lutea maintained for months by autografts of rat hypophyses, Endocrinology, 58:786–96, 1956.
28. Fisher, C., W. R. Ingram, and S. W. Ranson, *Diabetes Insipidus and the Neurohormonal Control of Water Balance.* Ann Arbor, Michigan, Edwards, 1938.
29. Flerkó, B., Zur hupothalamischen steuerung der gonadotrophen

Funktion der Hypophyse., Acta Morphol. Hung., 4:475–92, 1954.

30. ——, and G. Illei, Zur Frage der Spezifität des Einflusses von Sexualsteroiden auf hypothalamische Nervenstrukturen, Endokrinologie, 35:123–27, 1957.
31. ——, and J. Szentágothai, Oestrogen sensitive nervous structures in the hypothalamus, Acta Endocr., 26:121–27, 1957.
32. Foa, C., La greffe des ovaires en rélation avec quelques questions de biologie générale, Arch. Ital. Biol., 34:43–73, 1900.
33. Fujita, H., Electron microscope observation on neurosecretory granules in the pituitary posterior lobe of dog, Arch. Hist. Japon., 12:165–72, 1957.
34. Fulton, J. F., *A Bibliography of two Oxford Physiologists, Richard Lower and John Mayow*. London, Oxford University Press, 1935.
35. Ganong, W. F., Role of the nervous system in reproductive processes, in Cole, H. H., and P. T. Cupps, eds., *Reproduction in Domestic Animals*, 1:185. New York, Academic Press, 1959.
36. Gaunt, R., J. J. Chart, and A. A. Renzi, Endocrine pharmacology, Science, 133:613–21, 1961.
37. Gloor, P., Telencephalic influences upon the hypothalamus, in Field, W. S., R. Guillemin, and C. A. Carton, eds., *Hypothalamic-hypophysial Interrelationships*, pp. 74–113. Springfield, Thomas, 1956.
38. Gorski, R. A., and C. A. Barraclough, Differential effectiveness of small dosages of testosterone propionate in the induction of sterility in the female rat, Anat. Record, 139:304, 1961.
39. Green, J. D., Vessels and nerves of amphibian hypophyses: a study of the living circulation and of the histology of the hypophysial vessels and nerves, Anat. Record, 99:21–54, 1947.
40. ——, The comparative anatomy of the hypophysis, with special reference to its blood supply and innervation, Am. J. Anat., 88:225–312, 1951.
41. ——, C. D. Clemente, and J. De Groot, Rhinencephalic lesions and behavior in cats, J. Comp. Neurol., 108: 505–36, 1957.
42. ——, and G. W. Harris, The neurovascular link between the neurohypophysis and adenohypophysis, J. Endocrinol., 5:136–46, 1947.

43. ——, Observation of the hypophysioportal vessels of the living rat, J. Physiol., 108:359–61, 1949.
44. ——, and D. S. Maxwell, Comparative anatomy of the hypophysis and observations on the mechanism of neurosecretion, in Gorbman, A., ed., *Comparative Endocrinology*, pp. 368–92. New York, John Wiley, 1959.
45. ——, and V. L. van Breeman, Electron microscopy of the pituitary and observations on neurosecretion, Am. J. Anat., 97:177–228, 1955.
46. Guillemin, R., W. R. Hearn, W. R. Cheek, and D. E. Housholder, Control of corticotropin release: further studies with *in vitro* methods, Endocrinology, 60:488–506, 1957.
47. ——, and A. V. Schally, Re-evaluation of a technique of pituitary incubation *in vitro* as an assay for corticotropin releasing factor, Endocrinology, 65:555–62, 1959.
48. Haighton, J., An experimental enquiry concerning animal impregnation, Phil. Trans. Roy. Soc. London B., 87:159–96, 1797.
49. Harris, G. W., The induction of ovulation in the rabbit by electrical stimulation of the hypothalamo-hypophysial mechanism, Proc. Roy. Soc. (London) B., 122:374–94, 1937.
50. ——, *The Secreto-motor Innervation and Actions of the Neurohypophysis: an Investigation using the Method of Remote Control Stimulation*. Thesis for M.D. degree, Cambridge, University of Cambridge, 1944.
51. ——, Electrical stimulation of the hypothalamus and the mechanism of neural control of the adenohypophysis, J. Physiol., 107:418–29, 1948.
52. ——, Hypothalamo-hypophysial connexions in the Cetacea, J. Physiol., 111:361–67, 1950.
53. ——, Oestrous rhythm, pseudopregnancy and the pituitary stalk in the rat, J. Physiol., 111:347–60, 1950.
54. ——, The function of the pituitary stalk, Bull. Johns Hopkins Hosp., 97:358–75, 1955.
55. ——, Central control of pituitary secretion, in Field, J., H. W. Magoun, and V. E. Hall, eds., *Handbook of Physiology-Neurophysiology II*, p. 1012. American Physiological Society, Washington, D. C., 1960.
56. ——, The nervous system—follicular ripening, ovulation, and oestrous behaviour, in Lloyd, C. W., ed., *Endocrinology of Reproduction*, pp. 21–44. New York, Academic Press, 1959.

57. ——, The pituitary stalk and ovulation, in Villee, C. A., ed., *Control of Ovulation*. London, Pergamon, 1961.
58. ——, and D. Jacobsohn, Functional grafts of the anterior pituitary gland, Proc. Roy. Soc. (London) B., 139:263–76, 1952.
59. ——, and R. P. Michael, Hypothalamic mechanisms and the control of sexual behavior in the female cat, J. Physiol., 142: 26P, 1958.
60. ——, and J. W. Woods, The effect of electrical stimulation of the hypothalamus or pituitary gland on thyroid activity, J. Physiol., 143:246–74, 1958.
61. Haterius, H. O., and A. J. Derbyshire, Jr., Ovulation in the rabbit following upon stimulation of the hypothalamus, Am. J. Physiol., 119:329–30, 1937.
62. Heape, W., Ovulation and degeneration of ova in the rabbit, Proc. Roy. Soc. (London) B., 76:260–68, 1905.
63. Hebb, C. O., Chemical agents of the nervous system, Inter. Rev. Neurobiol., 1:165–93, 1959.
64. Houssay, B. A., A. Biasotti, and R. Sammartino, Modifications fonctionelles de l'hypophyse après les lésions infundibulo-tubériennes chez le crapaud, Compt. Rend. Soc. Biol., 120: 725–27, 1935.
65. Howell, W. H., The physiological effects of extracts of the hypophysis cerebri and infundibular body, J. Exp. Med., 3: 245–58, 1898.
66. Hume, D. M., and G. J. Wittenstein, The relationship of the hypothalamus to pituitary-adrenocortical function, in Mote, J. R., ed., *Proc. 1st Clinical ACTH Conference*, pp. 134–46. Philadelphia, Blakiston, 1950.
67. Jost, A., Problems of fetal endocrinology: the gonadal and hypophyseal hormones. Recent Prog. in Hormone Research, 8:379–413, 1953.
68. Klüver, H., Brain mechanisms and behavior with special reference to the rhinencephalon, J. Lancet, 72:567–74, 1952.
69. ——, and G. W. Bartelmez, Endometriosis in a rhesus monkey, Surg. Gynecol. Obstet., 92:650–60, 1951.
70. ——, and P. C. Bucy, Preliminary analysis of the function of the temporal lobes in monkeys, Arch. Neurol. Psychiat., 42:979–1000, 1939.
71. Kobayashi, H., and D. S. Farner, The effect of photoperiodic stimulation on phosphatase activity in the hypothalamo-hypophysial system of the white-crowned sparrow, *Zonotrichia*

Leucophrys Gambelli, Z. Zellforsch. Mikroskop. Anat., 53:-1-24, 1960.
72. Koikegami, H., T. Yamada, and K. Usui, Stimulation of amygdaloid nuclei and periamygdaloid cortex with special reference to its effect on uterine movements and ovulation, Folia Psychiat. et Neurol. Japon, 8:7–31, 1954.
73. Lister, M., in Lancaster, E., Ed., *The Correspondence of John Ray* (1675), p. 117. 1848.
74. Markee, J. E., J. W. Everett, and C. H. Sawyer, The relationship of the nervous system to the release of gonadotrophin and the regulation of the sex cycle, Recent Prog. in Hormone Research, 7:139–57, 1952.
75. ——, C. H. Sawyer, and W. H. Hollinshead, Activation of the anterior hypophysis by electrical stimulation in the rabbit, Endocrinology, 38:345–57, 1946.
76. ——, Adrenergic control of the release of luteinizing hormone from the hypophysis of the rabbit, Recent Prog. in Hormone Research, 2:117–31, 1948.
77. Marshall, F. H. A., Sexual periodicity and the causes which determine it, The Croonian Lecture. Phil. Trans. Roy. Soc. London B., 226:423–56, 1936.
78. ——, Exteroceptive factors in sexual periodicity, Biol. Rev., 17:68–90, 1942.
79. Martinez, C., and J. J. Bittner, A non-hypophysial sex difference in oestrous behaviour of mice bearing pituitary grafts, Proc. Soc. Exp. Biol. Med., 91:506–09, 1956.
80. Mason, J. W., The central nervous system regulation of ACTH secretion, in Jasper, H. H., L. D. Proctor, R. S. Nighton, W. C. Noshay, and R. T. Costello, eds., *Reticular Formation of the Brain*, pp. 645–62. Boston, Little, Brown, 1958.
81. McCann, S. M., and S. Taleisnik, Hypothalamic regulation of luteinizing hormone secretion, Science, 132:1496, 1960.
82. ——, The effect of a hypothalamic extract on the plasma luteinizing hormone (LH) activity of the oestrogenized, overiectomized rat, Endocrinology, 68:1071–73, 1961.
83. Michael, R. P., An investigation of the sensitivity of circumscribed neurological areas to hormonal stimulation by means of the application of oestrogens directly to the brain of the cat, in Kety, S. S. and J. Elkes, eds., *Regional Neurochemistry* Proc. 4th Int. Neurochem. Symp., pp. 465–79. London, Pergamon, 1961.

84. Nauta, W. J. H., and H. G. J. M. Kuypers, Some ascending pathways in the brain stem reticular formation, in Jasper, H. H., L. D. Proctor, R. S. Nighton, W. C. Noshay, and R. J. Costello, eds., *Reticular Formation of the Brain*, pp. 3–30. Boston, Little, Brown, 1958.
85. Nikitovitch-Winer, M., and J. W. Everett, Functional restitution of pituitary grafts re-transplanted from kidney to median eminence, Endocrinology, 63:916–30, 1958.
86. Oksche, A., D. F. Laws, F. I. Kamemoto, and D. S. Farner, The hypothalamo-hypophysial neurosecretory system of the white-crowned sparrow, *Zonotrichia Leucophrys Gambelli*, Z. Zellforsch Mikroskop. Anat., 51:1–42, 1959.
87. Oliver, G., and E. A. Schäfer, On the physiological action of extracts of pituitary body and certain other glandular organs, J. Physiol., 18:277–79, 1895.
88. Ott, I., and J. C. Scott, The action of infundibulin upon the mammary secretion, Proc. Soc. Exp. Biol. Med., 8:48–49, 1910.
89. Palay, S. L., The fine structure of the neurohypophysis, in Waelsch, H., ed., *Progress in Neurobiology*, II: 31–49. New York, Heober-Harper, 1957.
90. Pfeiffer, C. A. Sexual difference of the hypophyses and their determination by the gonads, Am. J. Anat., 58:195–225, 1936.
91. Popa, G. T., and U. Fielding, A portal circulation from the pituitary to the hypothalamic region, J. Anat. (London), 65: 88–91, 1930.
92. ——, Hypophysio-portal vessels and their colloid accompaniment, J. Anat. (London), 67:227–32, 1933.
93. Porter, R. W., The central nervous system and stress-induced eosinopenia, Recent Prog. in Hormone Research, 10:1–18, 1954.
94. Rasmussen, A. T., Innervation of the hypophysis, Endocrinology, 23:263–78, 1938.
95. Rolleston, H., *The Endocrine Organs in Health and Disease*, p. 12. London, Oxford Univ. Press, 1936.
96. Royce, P. C., and G. Sayers, Blood ACTH: effects of ether, pentobarbitol, epinephrine and pain, Endocrinology, 63:794–800, 1958.
97. Saffron, M. A., A. V. Schally and B. G. Benfey, Stimulation of the release of corticotropin from the adenohypophysis by a neurohypophysial factor, Endocrinology, 57:439–44, 1955.

98. Sawyer, C. H., Reproductive behavior, in Field, J., H. W. Magoun, and V. E. Hall, eds., *Handbook of Physiology-Neurophysiology II*, pp. 1225–40. Washington, D. C., American Physiological Society, 1960.
99. ——, and M. Kawakami, Interaction between the central nervous system and hormones influencing ovulation, in Villee, C. A., ed., *Control of Ovulation*, pp. 79–97. London, Pergamon, 1961.
100. Sayers, G., and P. Royce, Regulation of the secretory activity of the adrenal cortex, in Astwood, E. B., ed., *Clinical Endocrinology*, I:323–34. New York, Grune and Stratton, 1960.
101. Schally, A. V., R. N. Andersen, H. S. Lipscomb, J. M. Long, and R. Guillemin, Evidence for the existence of two corticotrophin-releasing factors, α and β, Nature, 188:1192–93, 1960.
102. ——, and R. Guillemin, Studies on corticotropin releasing factor: ion exchange chromatography of pituitary preparations, Texas Rep. Biol. Med., 18:133–46, 1960.
103. Scharrer, E., and B. Scharrer, Hormones produced by neurosecretory cells, Recent Prog. in Hormone Research, 10:183–232, 1954.
104. Schreiner, L., and A. Kling, Behavioral changes following rhinencephalic injury in cat, J. Neurophysiol., 16:643–59, 1953.
105. ——, Effects of castration on hypersexual behavior induced by rhinencephalic injury in cat, Arch. Neurol. Psychiat., 72:180–86, 1954.
106. ——, Rhinencephalon and behavior, Am. J. Physiol., 184:486–90, 1956.
107. Segal, S. J., and D. C. Johnson, Inductive influence of steroid hormone on the neural system: ovulation controlling mechanisms, Arch. Anat. Micr., 48 (Suppl.):261–74, 1959.
108. Selye, H., A syndrome produced by diverse nocuous agents, Nature, 138:32, 1936.
109. Shealy, C. N., and T. L. Peele, Studies on amygdaloid nucleus of cat, J. Neurophysiol., 20:125–39, 1957.
110. Singer, C., ed. and trans., *Vesalius on the Human Brain*. London, Oxford Univ. Press, 1952.
111. Smith, P. E., Postponed homotransplants of the hypophysis into the region of the median eminence in hypophysectomized male rats, Endocrinology, 68:130–43, 1961.
112. Taubenhaus, M., and S. Soskin, Release of luteinising hormone

from anterior hypophysis by an acetylcholine-like substance from the hypothalamic region, Endocrinology, 29:958–64, 1941.
113. Török, B., Lebendbeobachtung des Hypophysenkreislaufes an Hunden, Acta Morphol. Hung., 4:83–89, 1954.
114. Verney, E. B., The antidiuretic hormone and the factors which determine its release, Proc. Roy. Soc. (London) B., 135:25–106, 1947.
115. Von Euler, C., and B. Holmgren, The thyroxine 'receptor' of the thyroid-pituitary system, J. Physiol., 131:125–36, 1956.
116. Wingstrand, K. G., *The Structure and Development of the Avian Pituitary*, Sweden, Gleerup, Lund, 1951.
117. Wislocki, G. B., The vascular supply of the hypophysis cerebri of the cat, Anat. Record, 69:361–87, 1937.
118. ——, The vascular supply of the hypophysis cerebri of the Rhesus monkey and man, Research Publ. Assoc. Research Nervous Mental Disease, 17:48–68, 1938.
119. ——, and L. S. King, The permeability of the hypophysis and the hypothalamus to vital dyes, with a study of the hypophysial vascular supply, Am. J. Anat., 58:421–72, 1936.
120. Woods, J. W., and P. Bard, Thyroid activity in the chronic decerebrate cat with an isolated "island" of hypothalamus and pituitary, Federation Proc., 18:173, 1959.
121. ——, Antidiuretic hormone secretion in the cat with a centrally denervated hypothalamus, in *Proceedings of the First International Congress of Endocrinology*, Suppl. 51:113. Copenhagen, 1960.
122. Worthington, W. C., Some observations on the hypophyseal portal system in the living mouse, Bull. Johns Hopkins Hosp., 97:343–57, 1955.
123. Yamada, T., and M. A. Greer, The effect of bilateral ablation of the amygdala on endocrine function in the rat, Endocrinology, 66:565–74, 1960.

ROLF HASSLER

New Aspects of Brain Functions Revealed by Brain Diseases

IT IS APPROPRIATE that this discussion dealing with failures of the brain in disease should follow presentations on the performance of the normal brain. Is there, however, such a thing as a general failure of the brain, i.e., a brain-insufficiency, comparable to the insufficiency of the kidneys or of the heart? Even feeble-mindedness is not a general insufficiency of the brain, for simple motor or sensory functions may be unimpaired.

Sleep is, in many respects, a lowered functioning of the brain, but it is a normal, active switchover to repose in a well-coordinated manner (3, 52, 53, 56, 61) from which the brain can be aroused at any time. Neither is the pathological loss of consciousness a general insufficiency of the brain. For, under these conditions, the EEG shows large potential differences, (17, 18, 76, 80, 87), and in many narcotic states even potentials are evoked in an especially precise form in circumscribed fields of the cortex as a reaction to nonperceived sensory stimuli. The projection of the different sense organs upon the cortex can never be determined more clearly than in general barbiturate anesthesia (18,

ROLF HASSLER is Director of the Neuro-Anatomical Department of the Max Planck-Institut für Hirnforschung, Frankfurt am Main and Freiburg, Germany.

82, 118, 120) which is used to eliminate sensory perception during an operation. A complete silence of brain potentials occurs only in the exhaustion phase after grand mal seizures (63), or after air embolism of brain arteries, or in deep artificial hibernation (106). In this state there seems to be something like a general cerebral insufficiency, although it spares some of the brainstem centers.

The main deterrent to general brain insufficiency is that the brain contains a great number of organs which work together with a degree of intimacy. A failure of all these organs at the same time occurs only occasionally. In some brain diseases we are confronted with a functional loss of only one of these organs or of some of them which are shared by a system in an isolation which cannot yet be copied in experiment. Or there is a loss of cerebral performance at a level that does not exist in the animal, not even in those of highest development.

For these reasons one must be cautious in talking about brain insufficiency, as such. Instead, I would like to describe examples of special brain diseases which give us information about brain performance—information not yet yielded by animal experiments. On the other hand, the clinical neurologist is—in comparison to the experimental physiologist—in the awkward situation of having only one case appropriate to his investigation. This report will be based partly on clinical-pathological observations. Therefore, I would like to apologize in advance for the occasional subjectivity.

Somatotopic representations

As Hughlings Jackson (59) discovered in epileptic seizures starting from cortical lesions, there are representations in the cerebral cortex, not only of the contralateral

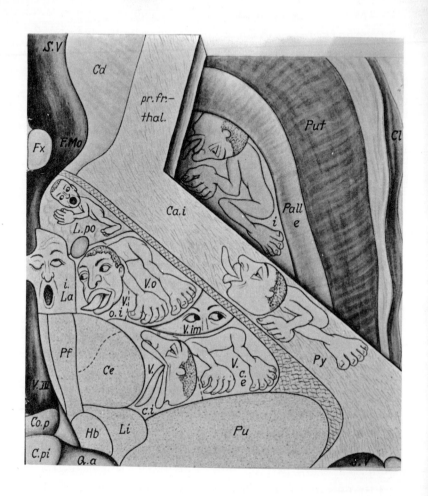

FIGURE 1

Scheme of functional localization within the ventral nuclei of the human thalamus, within the pallidum internum and the inner capsule. The different homunculi represent experiments with stimulation and coagulation in the human diencephalon during stereotactic brain operations. In the caudal ventral nuclei (V.c.e. and V.c.i. = VPL and VPM) the somatosensory homunculus is localized with the tongue as the most medial portion. The intermediate ventral nucleus (V.im) in front of the caudal ventral nuclei represents ipsiversive movements of the head or eyes. Moving frontally, the somatotopic representation of the afferents to the motor cortex follows in the oral ventral nuclei (V.o. and V.o.i. which together constitute the basal part of VL). The gaze movements elicited here are mostly contraversive. In the most rostral part of the ventral thalamus (nucleus latero-polaris = L.po, which corresponds to VA) is represented a complex motor and psychomotor effect, which resembles the stimulation effect of the supplementary motor area: sudden raising of the contralateral arm, contralateral deviation of the eyes, excited vocalization and impulse to speak. In a more medial position in the intralaminar nuclei (i.La), arousal and sleeping effects, symbolized by one open and one shut eye and by the yawning mouth, can be elicited. In the posterior part of the inner capsule the homunculus of the pyramidal tract is represented. Another homunculus of motor afferents lies in the internal segment of the pallidum. Consider the somatotopical arrangements from front to back in contrast to the arrangement of body regions from medial to lateral in the thalamic nuclei: Ca.i = Capsula interna; Cd = Caudatum; Ce = Centre médian; Cl = Claustrum; Co.p = Commissura posterior; C.pi = Corpus pineale; F.Mo = Foramen interventriculare Monroi; Fx = Fornix; Hb = Ganglion habenulae; i.La = intralaminar nuclei; Li = Limitans; L.po = Lateropolaris; Pall.e = Pallidum externum; Pall.i = Pallidum internum; Pf = Parafascicularis; pr.fr.-thal. = Fasciculi praefronto-thalamici; Pu = Pulvinar; Put = Putamen; Py = Tractus pyramidalis; Q.a = Corpus quadrigeminum anterius; S.V = Ventriculus lateralis; V.III = Ventriculus tertius; V.c.e = Ventrocaudalis externus; V.c.i = Ventrocaudalis internus; V.im = Ventro-intermedius; V.o = Ventro-oralis; V.o.i = Ventrooralis internus. (From Hassler, *Psychiatrie der Gegenwart,* in press) half of the body, but also of single regions of the body and parts of limbs. The great experience in cortical stimulation of conscious patients, obtained by Krause (70), and Foerster (14), as well as by Penfield (94) and Rasmussen (96), has revealed the details of the cortical representations of motility and of body surface. The single parts of the body are represented in a very different size according to their functional importance in the somatomotor and somatosensory cortex. Also, it is known from animal experiments that some regions of the body have very extensive representations in the cortex, while others have very small and circumscribed ones. This has been demonstrated especially in the cortical maps of Woolsey and his co-workers (82, 118, 120). Each of the different sensory systems, as well as the motor, has not only one but two or three cortical representations (97, 118), some of which are very well differentiated.

The same functional organization as that in the cortex can also be found in those thalamic nuclei which project upon the somatosensory cortex. Experiments with evoked potentials in animals indicate a representation of each single skin spot in the somatosensory nuclei of the diencephalon (88, 101). The clinical results of stimulations in patients suffering from intractable pain (43, 47, 48, 50) confirm this somatotopic sensory representation by subjective reports as well.

The motor homunculus in the cortex has at least one thalamic correspondent, the homunculus of the afferents to the pyramidal tract (37, 43), as demonstrated by our stimulations and coagulations in the rostral parts of the human thalamus (44, 47, 51). There are also clear indications of a representation of afferents to the supplementary motor area in the most rostral part of the ventral nuclei

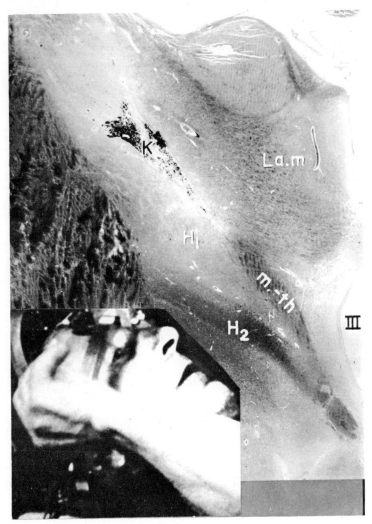

FIGURE 2

Brain slide prepared after Marchi-method showing the location of the coagulation focus (K) in the anterior thalamus in a case of postencephalitic Parkinsonism. The coagulation focus destroys nearly all of the left V.o.a which is demyelinated. The tip of the electrode lies in L.po near the Lamella medialis (La.m). Stimulation in this position resulted in a complex behavioral effect: the subject abruptly raising his contralateral arm and uttering unintelligible words in an excited manner, sometimes combined with a gaze deviation to the contralateral side. The H_2 bundle is preserved as is the Zona incerta above it, but the H_1 bundle is demyelinated and disappears in the base of the V.o.a. There are some Marchi degenerations in the mammillothalamic tract (m-th).

called VA or Lpo (Figure 1). In 15 percent of bipolar stimulations in the rostral ventral thalamus we saw approximately the same effect, a raising of the contralateral arm, contralateral deviation of the eyes and vocalization (47), as seen from the supplementary motor area by Penfield and Welch (97) (Figure 2).

There is also a representation of single body parts in the pallidum (47, 51). A similar locally different representation of single body regions can be assumed in other structures of the basal ganglia, such as the putamen and the substantia nigra; this assumption is based on clinicopathological findings in Parkinsonism (35) and in some forms of hyperkinesis. In the cerebellum the somatotopical localization has been experimentally established in carnivores (34) and in primates, but not clinically in the human. However, in the thalamic terminus of the fibers from the dentate nucleus, there is probably such a local representation in the human (47).

The complicated structures of the human brain are thus brought to life by these numerous homunculi in the deep cortical parts of the brain. This differentiation of the localized representations in different systems enriches our knowledge of the organization of the brain. It indicates that the brain is constructed according to the principle of chains of representations arranged one above the other. It possesses a vertical structure (39, 42, 45) with many vertically arranged chains of representations parallel to each other. It still remains to be discovered how far these representations of single body parts are able to compensate for each other under pathological conditions, and why each system has so many relays (Figure 3).

Integrating representations

In addition to these representations of single parts of the body or of muscle groups in different levels of the central nervous system and in addition to re-representations of the more simple ones, there are also, according to Jackson (59), highest level centers or integrating representations. For instance, there is the integration of all single somatosensory modalities which arise from an object recognized by palpation (73, 112, 116). In these integrating representations the sense impressions—such as the degrees of temperature, hardness and form of any object—can vanish completely but recognition of the object felt, such as a key or a matchbox, remains.

In all perceptions, especially if performed through one sense alone, the simple sensation must be supplemented by a motor action. This is most striking when trying to recognize something by its feel, when the visual sense is omitted. The contact of a spot of skin surface with an unusual object alone is not sufficient for the recognition of the consistency, material and form of that object. Movements of the palpating skin spots over the object are needed for such recognition. Without the active movements of the sensory surface, no integration of the single sense impressions occurs and no perception results (113, 41, 73). Thus, for perception, many sensory and some motor messages must be integrated.

In the case of the visual sense, the representations of the visual impressions of a recognized object are linked together by the performing of the movement of an active gaze. Each visual perception contains traces of the gaze movements which Von Holst and Mittelstädt (114) call "effer-

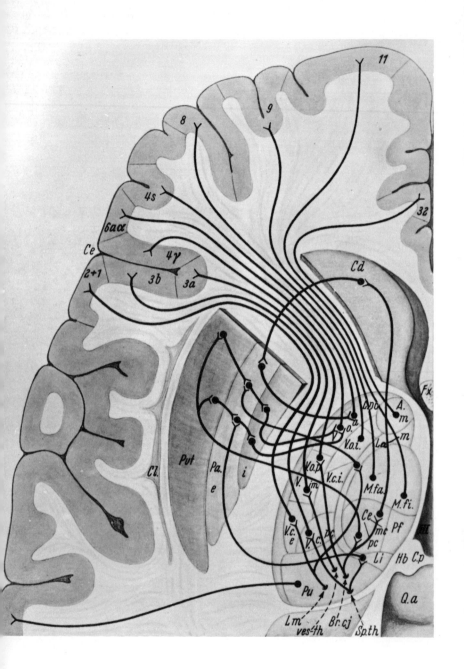

FIGURE 3

Diagram of the thalamo-cortical connections of the prefrontal lobe and central region of the human brain and of some of known afferent systems to the related thalamic nuclei. The cortical areas, according to Brodmann and Vogt, have dual conducting connections with special thalamic nuclei: V.c.e ≈ VPL—Area 1 + 2; V.c.pc ≈ VPI—area 3b; V.im—area 3a; V.o.p (posterior basal part of VL)—area 4 γ; V.o.a. (anterior basal part of VL)—area 6 a; L.po ($\approx$ VA)—area 6a β (supplementary area) + 4s; M.fa.p (part of dorsomedial nucleus)—area 46; M.fi.p—area 10; A.m—area 24. The afferent fiber chain to L.po originates from the emboliform nucleus and passes through the centre médian = ce (magno-cellular part: mc), the Caudatum (Cd) and the rostral part of the Pallidum externum. The afferent chain to V.o.a originates from the emboliform nucleus and passes through the parvocellular part of the centre médian (Ce.pc) and through the Putamen, Pallidum externum, Pallidum internum and thalamic fascicle (H_1). The afferent chain to V.o.p originates from the dentate nucleus and passes immediately to V.o.p (Br.cj). The afferent chain to V.im is the vestibulo-reticulo-thalamic tract (ves.th). The spino-thalamic tract (Sp.th) branches off collaterals to the Limitans (Li) before ending in the parvocellular part of the Ventro-caudal nucleus (V.c.pc). In the same manner, the Lemniscus medialis (Lm) branches off collaterals to the Lamella medialis (La.m). The latter and the limitans project to the pallidum externum. (From Hassler, Dtsch. med. Wschr., 1962).

ence-copies." The most instructive example of this is the subjective effect of acute paralysis of the muscle which turns the eye outward. The patient notices double images as soon as he looks in the direction of the paralyzed eye-muscle. If he glances in this direction he gets the impression that the double image is shifting in the contrary direction, although the eye with the paralyzed muscle has really made no movement at all. Even the central impulse for this gaze movement enters into the visual perception although no effective eye movement occurs in the periphery (41). A trace or a copy of the efferent gaze impulse is built into the process of visual perception. The role of the centrifugal impulses in the genesis of auditory perception has been investigated experimentally in connection with the discovery by G. Rasmussen (100) of the efferent cochlear bundle.

In many cases perception is restricted to extracting the meaningful from the less important details of the perceived object (41). For example, a complete half-sided loss of visual fields is compensated after some weeks in such a way that the patient can clearly perceive a whole cabbage lying in front of him, although in fact he sees nothing at all of one of its halves. Especially impressive is the restriction to the absorption of meaning in the process of fluent reading, in which many letters are not perceived at all.

Additionally, human beings possess brain structures which enable them to recognize certain symbols. These might even be some apparati which serve the sole purpose of recognizing those symbols developed by civilized man. There are some neurological patients who are not blind and who have not lost their ability to recognize objects and faces but nevertheless are unable to orient themselves or to find their way (67, 68). Other patients suddenly lose their ability to understand written and printed sentences and also the ability to recognize letters, although they are not blind or unable to recognize objects and spatial relations (67, 73, 91). They also are able to understand all that is spoken to them immediately and accurately.

An exceptionally clear case of such an inability to read, a pure alexia, which Beringer observed clinically (6), was caused by a softening of the base of the left occipital lobe by an embolic obstruction of a branch of the posterior cerebral artery as demonstrated in Figure 4. The patient, who survived a stroke for seven years, never regained the ability to read. She was able to recognize only a few letters immediately after reposal of visual fields by darkness, in spite of right side hemianopsia. If, for instance, the printed word "India" was offered to her she was unable to comprehend the word; but after glancing quickly over the letters several times she remarked that the word dealt in some way with elephants.* In spite of her reading deficit, she was able to comprehend as a Gestalt, the sphere of meaning to which this word referred, without deciphering any of the letters. This suggests that a lower level of perception exists. It does not depend on extracting the exact meaning of the sentence by recognition of special letters and words but on an undetermined comprehension of the sphere to which the sentence belongs. The comprehension of the

---

* Instead of "India" she read elephant and said literally: "It is still a foreign word, I knew at once that it must be something far away, in the tropics; it is hot, and so elephants came into my head. I always liked elephants so much." Instead of "fox" she read hare and said: "When first reading animal occurred to me, then it seemed to mean hare, then chicken, then hen."

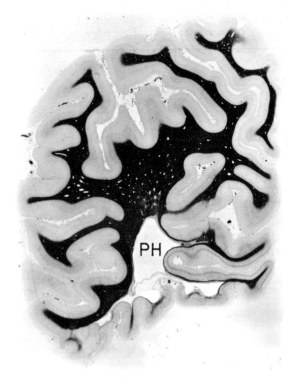

FIGURE 4

The left occipital lobe in Beringer's case of pure alexia, stained for fibers in the manner of Heidenhain-Woelcke. Because of an embolic occlusion of the left posterior cerebral artery, a "subcortical" softening of the white matter (PH) and the cortex occurred in the base of the occipital lobe, sparing parts of the superficial layers only. This softening is the pathological basis of the described form of pure alexia. The stripe of Gennari is preserved around the calcarine fissure in spite of the partial destruction of the optic radiation.

meaning by analysis of single letters and words, and their synthesis to a certain meaning is preceded by the absorption of the meaning from the less differentiated Gestalt, the so-called "sphere of discourse." The modern methods used to teach reading and writing by integral parts of sentences or words can be derived from this fact. The differentiated comprehension of the meaning of written or printed words depends on integrating brain structures in the base of the left occipital lobe in connection with some pulvinar nuclei (41, 45). It is not a function of a single sensory apparatus with its cortical representation.

Representations of self and affective behavior

From human cerebral pathology we further learn that beside the representations for recognition of visual and acoustic symbols there are also, in special parts of the brain, representations of the person himself. This is the somatic basis of the consciousness of self and self-reflection (4, 16, 36, 40, 66, 102). In compulsive and obsessive patients these higher representations of self seem to be overemphasized (5, 39). These patients are therefore unable to act easily because of frequently repeated reflections upon themselves, upon their intentions, and scruples about their actions and the possible consequences of their actions. Therefore these patients fail in the struggle of life if they have to be entirely dependent on themselves. They can be relieved unequivocally by psychosurgery of their compulsion to brood excessively over themselves and over their actions (5, 16, 39, 74, 84, 89). This can be done by bilateral undercutting of the medial convolutions of the orbital lobe (104) or by interruption of the connections with deeper brain structures (36, 39, 84, 121). After such operations the patients are no longer concerned with themselves but depend wholly on the inflow of stimuli.

The self-representations which are lost need not be conceived too abstractly. Each normal person can remember himself in past situations and can experience himself in the present under the special constellation of personal and social relationships. He is also able to represent himself anticipating a special situation which he plans or which may be caused by others. This projection of himself into the future is the basis of the most acute emotions of human experience, such as fear and anxiety. These future-related feelings are the moving forces in patients with compulsive states. The patients are unable to disregard and pass over the fear that if some action, which appears absurd to an observer, is omitted, an undefined misery will result. In spite of their nonsensical character, these excessively strong and insuperable inner representations of themselves in the future are an integrating feature of compulsive patients. In most of the severe cases, the self-aggression connected with them, which is often inhibited, cannot be improved by psychotherapy (5, 89). The specific therapy is—as mentioned—a bilateral interruption of certain fronto-thalamic systems which belong especially to the medial orbital lobe (36, 104) and to the medial surface of the frontal lobe (4). This may be a real and perhaps even a causal therapy, although personality changes occur which go beyond the relief of the compulsion. This indicates that the orbito-fronto-thalamic system is the system of representation of self and one's personal future. The restriction of the functional activity of this system results in reduction of anxiety, especially in patients with compulsive neurosis, as well as in other psychiatric patients with anxiety. This has been demonstrated by Freeman and Watts (16) and many others (5, 39, 84, 104, 121) through psychosurgical interventions in many types of psychoses.

In addition, some forms of affective behavior seem to be represented in the anterior thalamus near the lamella

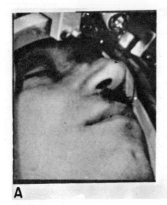 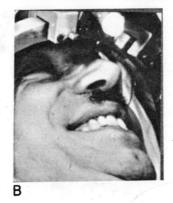

FIGURE 5

Stimulation of Pallidum internum (50/sec.) in a case of dystonia musculorum deformans results in a natural laughing with an adequate stimulus (B). This could be reproduced many times also when the unanesthetized, completely cooperative patient was asked not to laugh. (From Hassler, 47)

medullaris medialis (47). Four percent of our stimulations in this zone result in laughing with adequate effect, and 9 percent in smiling, beginning at the contralateral corner of the mouth (Figure 5).

The limbic lobe

As in the investigations of the frontal systems, stimulation and ablation of those higher structures of the brain which are thought to be connected with the sense of smell have also yielded no result in animal experiments. The neuronal systems above the prepiriform area, the primary cortical field of smell, occupied a silent region in respect of localization. This was especially true of the entorhinal area, the hippocampus, the fornix and mammillary bodies, and the anterior nuclei of the thalamus and their cortical projections in the cingulate gyrus. The system formed by these structures is called "limbic" by MacLean (78).

Wernicke (116) and Gamper (19) discovered that in alcoholics with Korsakoff-syndrome, the hypothalamic parts of this system, the mammillary bodies, are degenerated. There is an amnesia with disorientation in time and confabulation. This loss of cells of the mammillary bodies is related to the main symptom of Korsakoff-syndrome, the loss of recent memory. For a long time, however, the correlation between mammillary bodies and the ability to remember met with a strong skepticism.

Ten years ago cases of bilateral acute destruction, or softening of the hippocampus, were described by Grünthal (30), Glees and Griffith (21) and Ule (111) as the cause of a comatose state with loss of all speech and expression. Unfortunately Grünthal called this state "dementia," although a severe disorder of consciousness was present and the patient did not speak at all.

One can conclude from these pathological cases that the acute bilateral loss of the hippocampus results in a severe disorder of consciousness with disorientation. More recent experiments in this institute at UCLA have also shown that the function of these areas seems to be closely connected with memory function (11, 26, 27, 79).

The fibers which conduct the impulses from the hippocampus are concentrated in the fornix. A bilateral interruption of the anterior columns of the fornix is demonstrated in Figure 6. This illustrates the brain of a patient who had suffered temporal lobe seizures which ceased for ten months after the stereotactic coagulation of the right fornix without any undesirable effect (49). The patient lost her temporal EEG focus and was able to perform once more her profession as a librarian. Because of a recurrence of her seizures, interruption of the other fornix

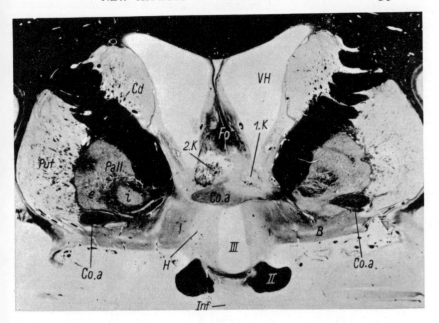

FIGURE 6

Bilateral stereotactic coagulation of the fornix (Fo) in front of the interventricular foramen with partial destruction of the anterior commissure. (Co.a) The coagulation of the right fornix (1.K) was performed one year before death; the coagulation of the left fornix (2.K), nine days before death. The demyelination of the first coagulation is complete in contrast to that of the left side. Note softening (H) in hypothalamus. (From Hassler and Riechert, 49)

was performed eleven months later. After this, Korsakoff-syndrome developed although her intelligence was preserved. She asked every person she met where she was and why she was there. She was unable to remember the right answers even for one minute and could not orient herself in time or situation. On the ninth day after the second fornicotomy she died from hyperthermia and a central breakdown of vegetative regulation.

The anatomical findings showed a softening in the

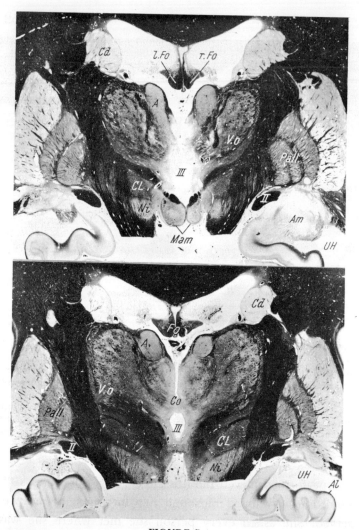

FIGURE 7

Transneuronal degeneration of the right mammillary body (Mam) with shrinkage and demyelination as an effect of the loss of afferent impulses during one year. On the other side only a small central demyelination in the mammillary body occurred nine days after the fornix interruption. The right fornix in the base of the lateral ventricle (r.Fo) has only a slight shrinkage in consequence of retrograde fiber degeneration. The picture below demonstrates the reduction in size of the right anterior nucleus (A) by transneuronal degeneration of a second order. (From Hassler and Riechert, 49) anterior hypothalamus (Figure 6) and a nondiagnosed glioma which was pressing on the hypothalamus frontally (Figure 9). The mammillary bodies were degenerated; a transneuronal degeneration of the second order was also present in the anterior nuclei of the thalamus (Figure 7). This unusual case indicates that bilateral destruction of the fornix produces in the human acute Korsakoff-syndrome with complete disorientation in time, space and situation, and inability to recall recent events.

The same case is also instructive with respect to the cause of temporal lobe epilepsy. This form of focal epilepsy, well-known since the investigations of Gibbs (20) and Jasper (60), is characterized by a specific kind of epileptic attack called psychomotor. The pathological substrata are focal lesions in the inner and frontal part of the temporal lobe surrounding the hippocampus, for instance in the amygdaloid nucleus (Figure 8). As Alfred Meyer and co-workers discovered (85, 86), these lesions can be produced by inflammation, by small angiomas, by nonexpanding gliomas or by brain cicatrix acquired during birth (Earle, Baldwin and Penfield, 10). The only important factor seems to be the close proximity of the pathological process to the hippocampus which has a low threshold for epileptic after-discharges, as has been shown by several workers (26, 28, 63, 79) and especially here at UCLA. Temporal lobe epilepsy is treated by extirpation of the temporal lobe with the EEG focus. To get a long lasting effect a large part of the hippocampus must then be removed, as pointed out by Falconer (11). Stimulation of the hippocampus before extirpation results in disorders of consciousness with a loss of attention. The bilateral extirpation of the temporal lobes, including most

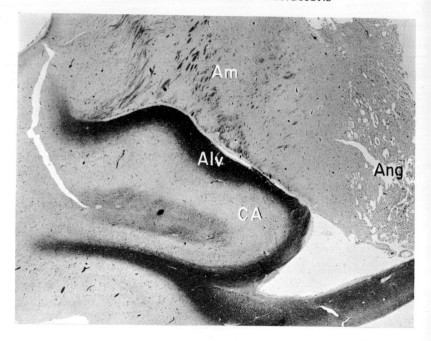

FIGURE 8

Calcifying angioma (Ang) of the right amygdaloid nucleus (Am) in a typical case of temporal lobe epilepsy published by Meyer-Mickeleit. The lateral parts of the amygdaloid nucleus are replaced by the tumor; the more medial parts are less damaged. Note that the hippocampus (CA) is not invaded by angioma in spite of the fact that a collateral injury has produced the seizures.

parts of the hippocampus, results in a loss of recent memory according to the results of neurosurgeons (11, 95). From all these experiences, however, it is still impossible to define the real function of the hippocampus. It is conceivable that it is important for the *temporal labeling* of experiences or a sort of activation of attention during perception.

In the case of bilateral fornicotomy just described, the substratum of the epilepsy was not situated in the temporal lobe but in the frontal part of the limbic lobe below the genu of the corpus callosum (Figure 9). It is well known that all structures of the limbic lobe have intimate relationships with the hippocampus, which is comparable to a common final pathway of the allocortical fields (78, 79).

Experiences of the symptoms which arise from lesions of higher parts of the limbic system in the human are scanty. Bilateral, therapeutic destructions of the cingulate gyri are effective against some schizophrenic symptoms, but the phenomena resulting from a bilateral functional loss of the cingulate gyri are not clear (74, 117). Therefore it is of interest to report what my associates and I could observe, after a bilateral stereotactic coagulation of both anterior nuclei of the thalamus in a schizophrenic patient who had suffered severe, crude, tactile and auditive hallucinations. The patient was indeed fully relieved of his worrying hallucinations, hallucinated needle pricks and other schizophrenic symptoms but at the same time a most severe amnestic syndrome appeared. The patient was completely disoriented in time, space and situation and was unable to recognize even his own family. This Korsakoff-syndrome disappeared so completely during the next twelve weeks that he could be discharged. Most of his schizophrenic symptoms did not reappear so that he has not needed treatment in a mental hospital since then, but his ability to remember has not completely recovered to a degree adequate to his age. Whitty and Lewin (117) observed a Korsakoff-syndrome after bilateral extirpation of the cingular gyri. In recent years a body of knowledge has accumulated about the limbic structures from experimental investigations in this and other institutes. Thus the

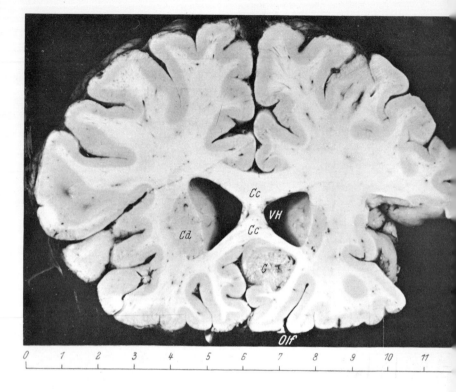

FIGURE 9

Coronal slice of the brain in the case of bilateral fornicotomy. The "temporal lobe epilepsy" is due to a small but expanding glioma with mucinous degeneration (G) in the right gyrus subcallosus. The right anterior horn of the lateral ventricle (VH) is not deformed and not displaced by the tumor. (From Hassler and Riechert, 49) scanty observations of human cases may contribute merely by adding the subjective experiences resulting from such disorders.

Localization of pathophysiological mechanisms

In the examples discussed, definite neurological and psychiatric symptoms could be referred to lesions of certain brain structures. Finding the seat of a disease, however, indicates only a little about the manner in which its symptoms develop. In a special group of nervous diseases the brain systems, the deficiency of whose substratum produces the clinical symptoms, have been detected. If a neuronal system working in conjunction with many others is lacking, the activity of the others is also disturbed. The other systems conduct pathological impulses which produce symptoms in the muscles or in sense organs. I would like to take as an example one of the most common of nervous diseases, Parkinsonism, or shaking palsy.

For a long time, the lesion of Parkinson's disease was assumed to be in the pallidum (75). The alterations described there, however, were not found in serial slices of many brains, especially in those of Parkinson patients who had died young (35). The alterations, formerly described as pathological, proved to be normal in old age (35). This also proves to be the case for the *état lacunaire* which was recently designated once again as the cause of Parkinsonism by Denny-Brown (9). Parkinsonism appeared as a sequela to the world-wide influenza epidemic of 1917 to 1918. In these cases, Tretiakoff (110), Foix (15), Luksch and Spatz (77) and others have found a circumscribed lesion in the substantia nigra. In Parkinson cases without preceding encephalitis, the deficit of these cells had been missed

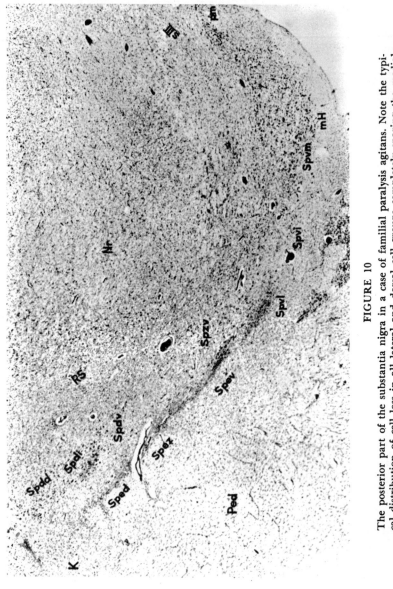

FIGURE 10

The posterior part of the substantia nigra in a case of familial paralysis agitans. Note the typical distribution of cell loss in all lateral and dorsal cell groups completely sparing the medial groups of smaller melanotic cells. (From Hassler, 35)

(107). According to the textbooks and handbooks the same symptoms could depend in some forms on cell deficit in the substantia nigra, in other forms on deficit in the pallidum (107).

An investigation of each aggregate of black pigmented cells in cases of Parkinsonism of hereditary or unknown cause constantly demonstrated (Figure 10) a severe destruction of specially grouped black, pigmented cells (29, 35, 42). The type of Parkinsonism caused by postencephalitis reveals in every case, including those with early onset, another pattern of loss of black cells (35) and a condensation of the neurofibrils in the scarce preserved cells (29, 33, 35). In Parkinsonism caused by heredity or by predisposition, the black nerve cells perish through argyrophilic and metachromatic spheric deposition in the plasma (29, 35, 42). A differentiation between these two most important forms of Parkinsonism can be made from the cytological alterations as well as by the pattern of distribution of the deficit of black cells. I cannot deal here with the lesions in other brain systems in cases of Parkinsonism (35).

How does such a deficit of black cells achieve the three main motor symptoms, rigidity, akinesis and tremor at rest? The black cells of the mesencephalon send out their processes, not ascending to the diencephalon as frequently assumed, but after crossing the midline in the commissura quadrigemina descending the bulb (40, 43) and spinal cord. This cell deficit has its effect at the level of the segmental motor nuclei. Consequently, those impulses which normally elicit the involuntary associated movements are lacking as are those impulses which bring about the many automatic movements that constitute an essential part of human motility (42, 44).

There is, however, in addition, a lack of the initiative necessary for the start of quick movements. In many normal types of quick movements an excitation of the small gamma-neurons precedes the innervation of the larger motoneurons, as Granit and others have found. In mammals the gamma-neurons are not able to elicit the contraction of the muscle itself, but they innervate the receptors (the muscle spindles) and make them more sensitive, so that these are forced to produce a contraction of the muscle via the monosynaptic reflex arc (22, 58, 72, 105). This setting in motion of the so-called external loop through the gamma-neurons is a trick of nature to spare energy for innervation and render possible a quick start like a self-starter (42). It is just this automatism that is lost by the Parkinson patients (44). This, at least to a great extent, is how the motor restriction and slowness is brought about.

To prove this conception we investigated the active reflex-reinforcement in Parkinson patients by the maneuver of Jendrassik. The latter reinforces the tendon-reflexes in normal persons by increasing the sensitivity of the muscle spindles via the gamma-fibers. These investigations showed that all forms of Parkinson's disease have a deficiency of reflex reinforcement, but in hemi-Parkinsonism (Figure 11) this deficiency occurs only on the rigid side (44, 71). In Parkinsonism, however, the peripheral gamma-neurons are not degenerated or absent but their rapid central innervation originating in the midbrain is impaired. The velocity of conduction in this pathway is 30 m/sec., according to Granit and Holmgren (23). The slow central innervation

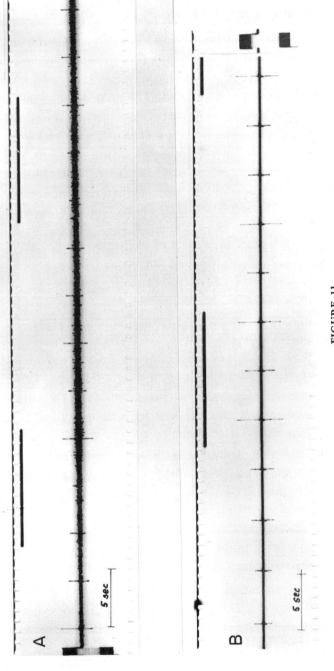

FIGURE 11

(A) Tendon reflexes of the rigid leg in a case of hemi-parkinsonism. The muscle activity increases during and after the reflex reinforcing maneuver of Jendrassik (mark). The amplitudes of tendon-reflexes decrease during this maneuver in contrast to the normal side. (B) Tendon reflexes of the same case but on the non-affected side are considerably increased during the same maneuver. No increase of muscle activity. (From Krienitz, 71) of the gamma-neurons and their excitability through skin stimuli is preserved in these patients. The tendon-reflexes are a little weaker, certainly not stronger in Parkinsonism, as would have to be postulated if a gamma-fiber spasm were the cause for the rigidity as is assumed by Schaltenbrand (103), Struppler (109), and Denny-Brown (9). As a result of the inability to quickly adjust the muscle spindles, the muscles lose their ability to keep the length constant, so that only the tension regulation is preserved (44, 113). On the other hand, the alpha-motoneurons are more excitable, as can be shown by eliciting electrically the monosynaptic reflex (Hoffman, 57; Paillard, 92; Schenck, 71). This does not have the effect of increased tendon-reflexes but only of increased slow stretch reflexes (13, 42), the so-called "myotatic reflexes". The myotatic reflexes affect mostly the constant muscle tension of Parkinson patients during wakefulness and the continuing electrical activity which can be recorded in the electromyogram. By slow stretch of the rigid muscles this increased tension is considerably enhanced (13, 71, 103). An increased contraction also occurs in antagonistic muscles in spite of their passive shortening. The increase of myotatic reflexes and their activity during wakefulness is an additional immediate consequence of the deficit of black cells in midbrain (42, 71). These, therefore, normally have an inhibiting action on the interneurons of myotatic reflexes (43).

As has been known for a long time, the rigidity ceases during sleep. This indicates the loss of an antagonistic influence of another system upon the myotatic reflexes. The system travels in the cortico-spinal pyramidal tract. There are two influences upon the myotatic reflex arc: a facilitating one from the pyramidal tract and an inhibiting one from the black cells in the midbrain (44, 46); normally the two are well balanced. If the inhibiting influence originating in the substantia nigra is absent, then the facilitating influence of the motor cortex prevails and an increase of muscle tension, namely rigidity, develops (Figure 12).

The same antagonism between the pigmented system and the pyramidal tract is operative for the symptom of tremor at rest. There is an old clinical observation, made even by Parkinson himself (93), that the tremor disappears after a hemiplegia. Bucy and Putnam took advantage of this for the therapy of tremor. Bucy (8) in 1938 was able to abolish the coarse tremor by extirpation of the motor cortex, including area 4 with the giant pyramidal cells. Putnam (99) achieved the same result in many cases by interruption of the pyramidal tract in the spinal cord as did Guiot and Pecker (32) in the cerebral peduncle. However, when the unavoidable postoperative one-sided weakness after pyramidotomy recovered, the tremor reappeared to the same degree. These operations have proved less effective for the relief of rigidity. The therapeutic success of these procedures indicates that an antagonistic balance indeed exists in the region of peripheral motor nuclei, and that this can be restored in Parkinson patients at the lower level.

Since the impulses of the pyramidal tract descending from the motor cortex are mainly initiated by afferent impulses originating in the specific subcortical projection nucleus (48), we assume that elimination of the thalamic projection nucleus would have the same effect against Parkinson symptoms but would have the advantage of sparing the weakness of limbs. We call the two structures in the thalamus "V.o.p." and "V.o.a." Together they form

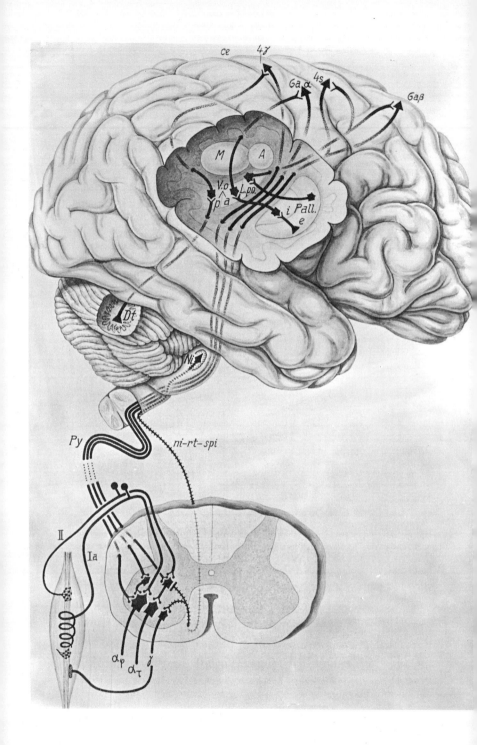

FIGURE 12

Schematic drawing of some parts of the extrapyramidal system with cortical connections to elucidate the pathophysiological mechanisms of Parkinson rigidity and their abolition by aimed subcortical coagulations. The rigidity is due to the cell loss of the substania nigra (Ni). The functional loss of the descending connections of the substantia nigra to the anterior horn, marked by cross lines, causes the inhibiting influence on the myotatic reflexes which are elicited by secondary endings (II) on the muscle spindles, specific interneurons and the tonic α-motoneurons (a_τ), to fail. Consequently, the myotatic reflexes and the muscle tonus are strongly increased. The myotatic reflexes are under the facilitating influence of some cortico-spinal systems. This influence can be reduced by extirpation of the motor cortex or by coagulation of the thalamic nuclei V.o.a and V.o.p, which project specifically to areas 4γ and $6a\alpha$. These operations restore the balance between the facilitating and inhibiting influences on the spinal mechanism of myotatic reflexes and abolish the rigidity. By coagulation of pallidum internum (Pall. i) the same effect can be obtained, because pallidum internum is the presynaptic neuron of V.o.a. The mechanism of monosynaptic tendon reflexes, which are mediated by the primary endings of muscle spindle and the phasic α-motoneurons (a_ϕ), is not disturbed in Parkinson's disease nor after the subcortical therapeutic coagulations. But the reflex reinforcement of the phasic tendon reflexes is no longer possible in Parkinson patients because the pathway for the fast central excitation of the γ-neurons from the substantia nigra is lost.

only the basal part of the VL, or ventral lateral, nucleus in the American nomenclature. Indeed, the coagulation of both structures produced, as expected, a complete relief from the tremor at rest without a paresis (48). Additionally, the operation also effected a complete relief from rigidity such as had never before been obtained. Since our first operation in 1952 this result has been confirmed by us in more than 500 Parkinson patients and by other teams as well (Figure 12).

These results would seem to prove that the facilitating influence of the pyramidal tract on rigidity and tremor is exerted from the afferent as well as from the efferent arc. The afferent arc originates from the basal part of the VL. This facilitating influence of the afferent arc to the pyramidal tract also operates from the neurons presynaptic to the V.o.a. which lie in the pallidum internum. The therapeutic results of Spiegel and Wycis (108), of Guiot (31), of Narabayashi (90) and of our team (44, 46, 48, 51) have demonstrated this fact. The fibers from the pallidum internum mostly terminate in the V.o.a. and the anterior basal part of VL (37). The effect of coagulation of the V.o.a. and the pallidum internum can be explained only by an ascending influence on the motor cortex (46). The rigidity can be abolished more specifically by a coagulation of the afferent arc of the pyramidal tract in the V.o.a. or the inner segment of the pallidum than by a coagulation of the efferent path, although no paresis occurs. After restoring the balance of the facilitating and inhibiting influences on the motor nuclei in this manner, the rigidity is abolished completely or almost completely, and tremor is much improved. The akinesis is, however, only slightly diminished. Even the reflex reinforcement evoked by the maneuver of Jendrassik is not restored by such an operation (71).

In these ways the pathogenesis of some Parkinson symptoms and their central conducting mechanism could be partly detected. They arise from an imbalance between facilitating pyramidal influences originating in the pallidum and dentate nucleus and the deficient inhibiting influences originating in the substantia nigra, which is destroyed in all Parkinson patients.

It is striking, however, that the additional elimination of the afferent inflow to the motor cortex seems not to produce additional symptoms of functional loss (44, 47). In the first week after unilateral elimination, one can observe only psychical symptoms together with a mimical facial paresis. These symptoms are increased fatigue, weakness of concentration, lack of spontaneity, unconcern and retardation of perception for only a few days. After unilateral coagulation these psychical changes relate to the elimination of the pallidum internum and the V.o.a. Further, after symmetrical elimination of these systems only psychical symptoms of functional loss result, but they are more severe and recovery is slower. There may be retardation of thinking, unconcern for associates, tactlessness, lack of interest, inability to concentrate, lack of energy, loss of imagination, lack of initiative and a severe disorder of memory, which is due mostly to a deficiency in attention and to dullness. Some of these symptoms may be due to accidental coagulation of fronto-thalamic pathways.

Therefore the internal segment of the pallidum, probably as well as the external segment, seems to be not a motor, but a psychomotor center or activation mechanism (43, 44, 46, 47). Stimulation of the pallidum results in a pseudo-arousal effect in behavior even in patients who are in a general anesthesia. The pupils dilate, the eyes open, the gaze moves around, the patients get a little reactive and say something, although they are still disoriented. The pallidum would thus appear to belong to the reticular activating system of Moruzzi and Magoun as a rostral continuation. With respect to the fiber connections with the thalamus the pallidum is included in the unspecific projection system to the cerebral cortex. The pallidum seems to be a very important link in the general activating system for the cerebral cortex. Its structure in fiber and cell preparations is typically reticular, more so than in the brain stem regions that are so named.

Both the pathological substratum in the cell loss of the substantia nigra and the normal fiber systems, through which the consequences of functional loss develop have now been found for the Parkinson syndrome. By the interruption of such fiber systems it is now possible to suppress the distressing symptoms, although the pathological process is not altered. Only the knowledge of both substrata can explain all phenomena of such a disease and also the rational treatment. The collaboration of clinical neurologists, neuropathologists, neurophysiologists and neurosurgeons will provide us with similar knowledge of other nervous diseases.

This survey which I have dared to give you has been very subjective and necessarily very incomplete. Its intention has been to point out the possible contributions of clinical pathological cases to brain research. In many places these clinical findings could later be checked by such experiments as, for example, those done by Peterson, Magoun and Lindsley in the case of tremor (98). This has not been possible with other disorders, either because they are primarily human diseases or because the animal is unable to give differentiated information about its inner experiences. In spite of the steady and rapid improvement of the experimental technique, the neurological clinic will continue to give a wealth of suggestions for experimental brain research, for cybernetics and for information theory.

REFERENCES

1. Adrian, E. D., Double representation of the feet in the sensory cortex of the cat, J. Physiol., 98:16P–18P, 1940.
2. ——, Afferent discharges to the cerebral cortex from peripheral sense organs, J. Physiol., 100:159–91, 1941.
3. Akert, K., W. P. Koella, and R. Hess, Jr., Sleep produced by electrical stimulation of the thalamus, Amer. J. Physiol., 168: 260–67, 1952.
4. Beringer, K., Über Störungen des Antriebes bei einem von der unteren Falxkante ausgehenden doppelseitigen Meningeom, Z. Neur., 171:451–74, 1941.
5. ——, Zur Frage der Leukotomie, Med. Klin., 44:853–57, 1949.
6. ——, and J. Stein, Analyse eines Falles von "reiner" Alexie, Z. Neur., 123:472–78, 1930.
7. Bernhard, C. G., and E. Bohm, Cortical representation and functional significance of the corticomotorneuronal system, Arch. Neurol. Psychiat., 72:473–502, 1954.
8. Bucy, P. C., and T. J. Case, Tremor: Physiological mechanisms and abolition by surgical means, Arch. Neurol. Psychiat., 41: 721–46, 1939.
9. Denny-Brown, D., Diseases of the basal ganglia, Their relation to disorders of movement, Lancet, I:1099–1105, 1155–62, 1960.
10. Earle, K. M., M. Baldwin, and W. Penfield, Incisural sclerosis and temporal lobe seizures produced by hippocampal herniation at birth, Arch. Neurol. Psychiat., 69:27–42, 1953.
11. Falconer, M. A., D. Hill, A. Meyer, W. Mitchell, and D. A. Pond, Treatment of temporal lobe epilepsy by temporal lobectomy, Lancet, 1:827–35, 1955.

12. Fénelon, F., and F. Thiébaut, Résultats du traitement neurochirurgical d'une rigidité Parkinsonienne par intervention strio-pallidale unilatérale, Revue Neurol., 83:280–83, 1950.
13. Foerster, O., Schlaffe und spastiche Lähmung, in Bethe, A., O. Bergmann, G. Embden and A. Ellinger, eds., *Handbuch der normalen und pathologischen Physiologie*, 10:893–972. Berlin, Springer, 1927.
14. ——, Motorische Felder und Bahnen, in Bumke, O., and O. Foerster, eds., *Handbuch der Neurologie*, 6:1–357. Berlin, Springer, 1936.
15. Foix, Ch., Les lésions anatomiques de la maladie de Parkinson, Revue Neurol., 28:593, 1921.
16. Freeman, W., and J. W. Watts, *Psychosurgery, in the Treatment of Mental Disorders and Intractable Pain*, 2d ed. Springfield, Thomas, 1950.
17. French, J. D., Brain lesions associated with prolonged unconsciousness, Arch. Neurol. Psychiat., 68:727–40, 1952.
18. ——, M. Verzeano, and H. W. Magoun, A neural basis of the anesthetic state, Arch. Neurol. Psychiat., 69:519–29, 1953.
19. Gamper, E., Zur Frage der Polioencephalitis haemorrhagica der chronischen Alkoholiker, Dtsch. Z. Nervenheilk., 102:122–29, 1928.
20. Gibbs, E. L., F. A. Gibbs, and B. Fuster, Psychomotor epilepsy, Arch. Neurol. Psychiat., 60:331–39, 1948.
21. Glees, P., and H. B. Griffith, Bilateral destruction of the Hippocampus (Cornu Ammonis) in a case of dementia, Mschr. Psychiatr., 123:193–204, 1952.
22. Granit, R., *Receptors and Sensory Perception*. New Haven, Yale Univ. Press, 1955.
23. ——, and B. Holmgren, Two pathways from brainstem to gamma ventral horn cells, Acta Physiol. Scand., 35:93–108, 1955.
24. ——, B. Holmgren, and P. A. Merton, The two routes for excitation of muscle and their subservience to the cerebellum, J. Physiol., 130:213–24, 1955.
25. Grantham, E. G., and R. G. Spurling, Selective lobotomy in the treatment of intractable pain, Ann. Surg., 137:602–08, 1953.
26. Green, J. D., and W. R. Adey, Electrophysiological studies of hippocampal connections and excitability, EEG Clin. Neurophysiol., 8:245–62, 1956.

27. ——, C. D. Clemente, and J. De Groot, Rhinencephalic lesions and behavior in cats. An analysis of the Klüver-Bucy syndrome with particular reference to normal and abnormal sexual behavior, J. Comp. Neurol., 108:505–45, 1957.
28. ——, and T. Shimamoto, Hippocampal seizures and their propagation, Arch. Neurol. Psychiat., 70:687–702, 1953.
29. Greenfield, J. G., and F. D. Bosanquet, The brain stem lesion in Parkinsonism, J. Neurol. Neurosurg. Psychiat., 16:213–26, 1953.
30. Grünthal, E., Über das klinische Bild nach umschriebenem beiderseitigem Ausfall der Ammonshornrinde. Ein Beitrag zur Kenntnis der Funktion des Ammonshorns, Mschr. Psychiatr., 113:1–16, 1947.
31. Guiot, G., and S. Brion, Traitement des mouvements anormaux par la coagulation pallidale, Technique et résultats, Revue Neurol., 89:578–80, 1953.
32. ——, and J. Pecker, Tractotomie mésencéphalique antérieure pour tremblement Parkinsonien, Revue Neurol., 81:387–88, 1949.
33. Hallervorden, J., Anatomische Untersuchungen zur Pathogenese des postencephalitischen Parkinsonismus, Dtsch. Z. Nervenheilk., 136:68–77, 1935.
34. Hampson, J. L., Relationships between cat cerebral and cerebellar cortices, J. Neurophysiol., 12:37–50, 1949.
35. Hassler, R., Zur Pathologie der Paralysis agitans und des postenzephalitischen Parkinsonismus, J. Psychol. Neurol., 48:387–476, 1938.
36. ——, Über die Thalamus-Stirnhirn-Verbindungen beim Menschen, Nervenärzt., 19:9–12, 1948.
37. ——, Über die afferenten Bahnen und Thalamuskerne des motorischen Systems des Grosshirns, I and II, Arch. Psychiatr., 182:759–818, 1949.
38. ——, Über die afferente Leitung und Steuerung des striären Systems, Nervenärzt, 20:537–41, 1949.
39. ——, Über die anatomischen Grundlagen der Leukotomie, Fortschr. Neurol. Psychiat., 18:351–67, 1950.
40. ——, Contribution morphologique à la physiologie des lobes frontaux, in *Compte Rendu I, Congrès Mondial de la Psychiatrie*, III:118–27. Paris, Hermann, 1950.
41. ——, Über die bei der optischen Wahrnehmung beteiligten Hirnsysteme, Zentralbl. Neurol., 112:152, 1951.

42. ——, Extrapyramidal-motorische Syndrome und Erkrankungen, in *Handbuch Inneren Medizin*, V(Pt. 3):676–904. Heidelberg, Springer, 1953.
43. ——, Functional anatomy of the thalamus, *Acta y Trabajos VI Congress Latinoamericano Neurocir.*, pp. 754–87. Montevideo, 1955.
44. ——, Die extrapyramidalen Rindensysteme und die zentrale Regelung der Motorik, Dtsch. Z. Nervenheilk., 175:233–58, 1956.
45. ——, Anatomy of the Thalamus, in Schaltenbrand, G., and P. Bailey, eds., *Introduction to Stereotaxis with an Atlas of the Human Brain*, 1:230–90. Stuttgart, Thieme, 1959.
46. ——, Stereotactic brain surgery for extrapyramidal motor disturbances, in Schaltenbrand, G., and P. Bailey, eds., *Introduction to Stereotaxis with an Atlas of the Human Brain*, 1:472–88. Stuttgart, Thieme, 1959.
47. ——, Motorische und sensible Effekte umschriebener Reizungen und Ausschaltungen im menschlichen Zwischenhirn, Dtsch. Z. Nervenheilk., 183:148–71, 1961.
48. ——, and T. Riechert, Indikationen und Lokalisationsmethode der gezielten Hirnoperationen, Nervenärzt, 25:441–47, 1954.
49. ——, Über einen Fall von doppelseitiger Fornicotomie bei sogenannter temporaler Epilepsie, Acta Neurochir., 5:330–40, 1952.
50. ——, Klinische und Anatomische Befunde bei stereotaktischen Schmerzoperationen im Thalamus, Arch. f. Psychiatr., 200:93–122, 1959.
51. ——, T. Riechert, F. Mundinger, W. Umbach, and J. A. Ganglberger, Physiological observations in stereotaxic operations in extrapyramidal motor disturbances, Brain, 83:337–50, 1960.
52. Hess, W. R., Hirnreizversuche über den Mechanismus des Schlafes, Arch. f. Psychiatr., 86:287–92, 1929.
53. ——, Das Schlafsyndrom als Folge diencephaler Reizung, Helvet. Physiol. Acta, 2:305–44, 1944.
54. ——, Korrespondierende Symptome aus Stirnhirn, Innerer Kapsel und vorderem Thalamus, Helvet. Physiol. Acta, 6:731–38, 1948.
55. ——, *Das Zwischenhirn*, 2d ed. Basel, Benno Schwabe, 1954.
56. Hess, Jr., R., W. P. Koella, and K. Akert, Cortical and sub-

cortical recordings in natural and artificially induced sleep in cats, EEG Clin. Neurophysiol., 5:75–90, 1953.
57. Hoffmann, P., Die Beziehungen des Willens und der einfachsten Reflexformen zueinander, Arch. f. Psychiatr., 185:736–42, 1950.
58. ——, Die Aufklärung der Wirkung des Jendrassikschen Handgriffs durch die Arbeiten von Sommer und Kuffler, Dtsch. Z. Nervenheilk., 166:60–64, 1951.
59. Jackson, J. H., in Taylor, J., ed., *Selected Writings*, Vols. I and II. New York, Basic Books, 1958.
60. Jasper, H. H., and J. Kershman, Electroencephalographic classification of the epilepsies, Arch. Neurol. Psychiat., 45:903–43, 1941.
61. Jouvet, M., F. Michel, and D. Mounier, Analyse électroencéphalographique comparée du sommeil physiologique chez le chat et chez l'homme, Revue Neurol., 103:189–205, 1960.
62. Jung, R., Physiologische Untersuchungen über den Parkinsontremor und andere Zitterformen beim Menschen, Z. Neur., 173:263–332, 1941.
63. ——, Hirnelektrische Untersuchungen über den Elektrokrampf, Arch. f. Psychiatr., 183:206–44, 1949.
64. ——, and R. Hassler, Extrapyramidal motor system, in Field, J., H. W. Magoun, V. E. Hall, eds., *Handbook of Physiology-Neurophysiology*, II:863–927. Washington, D. C., American Physiological Society, 1960.
65. Kaada, B. R., Somato-motor, autonomic and electrocorticographic responses to electrical stimulation of "rhinencephalic" and other structures in primates, cat, and dog, Acta Physiol. Scand., 24(Suppl. 83), 1951.
66. Kleist, K., Gehirnpathologische und lokalisatorische Ergebnisse V. Das Stirnhirn im engeren Sinne und seine Störungen, Z. Neur., 131:442–52, 1930.
67. ——, *Gehirnpathologie*. Leipzig, Joh. Ambr. Barth, 1934.
68. ——, Über Form—und Ortsblindheit bei Verletzung des Hinterhauptlappens, Dtsch. Z. Nervenheilk., 138:206–14, 1935.
69. Klüver, H., and P. C. Bucy, Preliminary analysis of functions of the temporal lobes in monkeys, Arch. Neurol. Psychiat., 42:979–1000, 1938.
70. Krause, F., *Surgery of the Brain and Spinal Cord Based on Personal Experiences*. 3 vols., New York, Rebman, 1909–1912.

71. Krienitz, E., Elektromyographische Untersuchungen über die eigenreflektorische Erregbarkeit der rigiden Muskalatur bei Parkinsonkranken, Freiburg, Dissertation, 1961.
72. Kuffler, S. W., C. C. Hunt, and J. P. Quilliam, Function of medullated small-nerve fibers in mammalian ventral roots: Efferent muscle spindle innervation, J. Neurophysiol., 14:29–54, 1951.
73. Lange, J., Agnosien und Apraxien, in *Handbuch der Neurologie*, 6:807–960. Berlin, Springer, 1936.
74. Le Beau, J., M. Choppy, J. Gaches, and M. Rosier, *Psychochirurgie et fonctions mentales: techniques, Résultats, Applications physiologiques*. Paris, Masson, 1954.
75. Lewy, F. H., *Die Lehre vom Tonus und der Bewegung*. Berlin, Springer, 1923.
76. Lindsley, D. B., L. H. Schreiner, W. B. Knowles, H. W. Magoun, Behavioral and EEG changes following chronic brainstem lesions in the cat, EEG Clin. Neurophysiol., 2:483–98, 1950.
77. Luksch, F., and H. Spatz, Die Veränderungen im Zentralnervensystem bei Parkinsonismus in den Spätstadien der Encephalitis epidemica, Münch. med. Wschr., 70:1245–47, 1923.
78. MacLean, P. D., Some psychiatric implications of physiological studies on frontotemporal portion of limbic system (visceral brain), EEG Clin. Neurophysiol., 4:407–18, 1952.
79. ——, Chemical and electrical stimulation of hippocampus in unrestrained animals. II. Behavioral findings, Arch. Neurol. Psychiat., 78:128–42, 1957.
80. Magoun, H. W., *The Waking Brain*. Springfield, Thomas, 1958.
81. Malamud, N., and S. A. Skillicorn, Relationship between the Wernicke and the Korsakoff syndrome, Arch. Neurol. Psychiat., 76:585–96, 1956.
82. Marshall, W. H., C. N. Woolsey, and Ph. Pard, Observations on cortical somatic sensory mechanisms of cat and monkey, J. Neurophysiol., 4:1–24, 1941.
83. Melzack, R., W. A. Stotler, and W. K. Livingston, Effects of discrete brain stem lesions in cats on perception of noxious stimulation, J. Neurophysiol., 21:353–67, 1958.
84. Meyer, A., and E. Beck, *Prefrontal Leucotomy and Related Operations. Anatomical Aspects of Success and Failure*. Edinburgh, Oliver and Boyd, 1954.

85. ——, The hippocampal formation in temporal lobe epilepsy, Proc. Roy. Soc. Med., 48:457–62, 1955.
86. ——, M. A. Falconer, and E. Beck, Pathological findings in temporal lobe epilepsy, J. Neurol. Neurosurg., 17:276–85, 1954.
87. Moruzzi, G., and H. W. Magoun, Brain stem reticular formation and activation of the EEG, EEG Clin. Neurophysiol., 1: 455–73, 1949.
88. Mountcastle, V. B., and E. Henneman, The representations of tactile sensibility in the thalamus of the monkey, J. Comp. Neurol., 97:409–40, 1952.
89. Müller, M., Über die präfrontale Leukotomie unter besonderer Berücksichtigung eines Falles von Zwangskrankheit, Nervenärzt, 19:97–107, 1948.
90. Narabayashi, H., and T. Okuma, Procaine-oil blocking of the globus pallidus for the treatment of rigidity and tremor of Parkinsonism, Proc. Jap. Acad., 29:134–37, 1953.
91. Nielsen, J. M., *Agnosia, Apraxia, Aphasia. Their Value in Cerebral Localization*, 2d ed. New York, Hoeber, 1948.
92. Paillard, J., *Réflexes et régulations d'origine proprioceptive chez l'homme.* Paris, Librairie Arnette, 1955.
93. Parkinson, J., An essay on the shaking palsy. London (1817), Reprinted: Arch. Neurol. Psychiat., 7:681–710, 1922.
94. Penfield, W., and H. H. Jasper, *Epilepsy and the Functional Anatomy of the Human Brain.* Boston, Little, Brown, 1954.
95. ——, and B. Milner, Memory deficit produced by bilateral lesions in the hippocampal zone, Arch. Neurol. Psychiat., 79:475–97, 1958.
96. ——, and Th. Rasmussen, *The Cerebral Cortex of Man.* New York, Macmillan, 1950.
97. ——, and K. Welch, The supplementary motor area of the cerebral cortex, Arch. Neurol. Psychiat., 66:289–317, 1951.
98. Peterson, E. W., H. W. Magoun, W. S. McCulloch, and D. B. Lindsley, Production of postural tremor, J. Neurophysiol., 12: 371–84, 1949.
99. Putnam, T. J., Treatment of unilateral paralysis agitans by section of the lateral pyramidal tract, Neurol. Psychiat., 44: 950–76, 1940.
100. Rasmussen, G. L., Further observation of the efferent cochlear bundle, J. Comp. Neurol., 99:61–74, 1953.

101. Rose, J. E., and V. B. Mountcastle, The thalamic tactile region in rabbit and cat, J. Comp. Neurol., 97:441–90, 1952.
102. Rylander, G., Personality changes after operations on the frontal lobes. A clinical study of 32 cases, Acta Psychiat. Neurol., (Suppl. XX) 1939.
103. Schaltenbrand, G., and H. J. Hufschmidt, The role of the pyramidal systems in the organization of motility, in Schaltenbrand, G., and P. Bailey, eds. *Introduction to Stereotaxis with an Atlas of the Human Brain,* I:354–71. Stuttgart, Thieme, 1959.
104. Scoville, W. B., Selective cortical undercutting as a means of modifying and studying frontal lobe function in man, J. Neurosurg., 6:65–73, 1949.
105. Sommer, H., Periphere Bahnung von Muskeleigenreflexen als Wesen des Jendrassikschen Phänomens, Dtsch. Z. Nervenheilk., 150:249–62, 1940.
106. Soriano, V., and C. B. de Soriano, Clinical and experimental studies of the electrical silence of the cortex. Abstr. VII. Internat. Congr. Neurology Rom., Excerp. Med. Int. Congr. Ser., 38:69–70. Amsterdam, 1961.
107. Spatz, H., Physiologie und Pathologie der Stammganglien, in Bethe, A., O. Bergmann, G. Embden, and A. Ellinger, eds., *Handbuch der normalen und pathologischen Physiologie,* 10:318–417. Berlin, Springer, 1927.
108. Spiegel, E. A., and H. T. Wycis, Ansotomy in Paralysis agitans, Arch. Neurol. Psychiat., 71:598–614, 1954.
109. Struppler, A., and R. Preuss, Untersuchungen über periphere und zentrale Faktoren der Eigenreflexerregbarkeit am Menschen mit Hilfe des Jendrassikschen Handgriffes, Pflügers Arch., 268:425–34, 1959.
110. Trétiakoff, C., Contribution à l'étude de l'anatomie pathologique de locus niger de Soemmering avec quelques déductions rélatives à la pathogénie des trouble du tonus musculaire de la maladie de Parkinson, Thèse de Paris, 1919.
111. Ule, G., Korsakow-Psychose nach doppelseitiger Ammonshornzerstörung mit transneuronaler Degeneration der Corpora mammillaria, Dtsch. Z. Nervenheilk., 165:446–56, 1951.
112. von Hattingberg, I., Sensibilitätsuntersuchungen an Kranken mit Schwellenverfahren. *Sitz. Ber. Heidelberger Akad. Wiss. Math. Nat. Kl.* 10. Heidelberg, Abhandlung, 1939.

113. von Holst, E., Zentralnervensystem und Peripherie in ihrem gegenseitigen Verhältnis, Klin. Wschr., 29:97–105, 1951.
114. ——, and H. Mittelstaedt, Das Reafferenzprinzip, Naturwiss., 37:464–76, 1950.
115. Walker, A. E., Cerebral pedunculotomy II. Parkinsonian tremor, J. Nerv. Ment. Dis., 116:766–75, 1952.
116. Wernicke, C., *Lehrbuch der Gehirnkrankheiten* I and III. Berlin, Fischer, 1881 and 1883.
117. Whitty, C. W. M., and W. Lewin, A Korsakoff syndrome in the post-cingulectomy confusional state, Brain, 83:648–53, 1960.
118. Woolsey, C. N., Organization of somatic sensory and motor areas of the cerebral cortex, in Harlow, H. F., and C. N. Woolsey, eds., *Biological and Biochemical Bases of Behavior*, pp. 63–81. Madison, Univ. of Wisconsin Press, 1958.
119. ——, and D. Fairman, Contralateral, ipsilateral and bilateral representation of cutaneous receptors in somatic areas I and II of the cerebral cortex of pig, sheep and other mammals, Surgery, 19:684–702, 1946.
120. ——, W. H. Marshall, and P. Bard, Representation of cutaneous tactile sensibility in the cerebral cortex of the monkey as indicated by evoked potentials, Bull. Johns Hopkins Hosp., 70:399–441, 1942.
121. Yakovlev, P. I., H. Hamlin, and W. Sweet, Frontal lobotomy: neuroanatomical observations, J. Neuropath., Exp. Neurol., 9:250–85, 1950.